Lecture Notes in Control and Information Sciences 391

Editors: M. Thoma, F. Allgöwer, M. Morari

Ali Khaki-Sedigh and Bijan Moaveni

Control Configuration Selection for Multivariable Plants

Springer

Authors

Prof. Ali Khaki-Sedigh

Faculty of Electrical and Computer Engineering
Department of Control Engineering
K. N. Toosi University of Technology
Tehran, 14317-14191
Iran
E-mail: sedigh@kntu.ac.ir

Dr. Bijan Moaveni

School of Railway Engineering
Iran University of Science and Technology
Tehran, 16846-13114
Iran
E-mail: b_moaveni@iust.ac.ir

ISBN 978-3-642-03192-2 e-ISBN 978-3-642-03193-9

DOI 10.1007/978-3-642-03193-9

Lecture Notes in Control and Information Sciences ISSN 0170-8643

Library of Congress Control Number: Applied for

Typeset & Cover Design: Scientific Publishing Services Pvt. Ltd., Chennai, India.

Printed in acid-free paper

5 4 3 2 1 0

springer.com

Preface

Control of multivariable industrial plants and processes has been a challenging and fascinating task for researchers in this field. The analysis and design methodologies for multivariable plants can be categorized as centralized and decentralized design strategies. Despite the remarkable theoretical achievements in centralized multivariable control, decentralized control is still widely used in many industrial plants. This trend in the beginning of the third millennium is still there and it will be with us for the foreseeable future. This is mainly because of the easy implementation, maintenance, tuning, and robust behavior in the face of fault and model uncertainties, which is reported with the vast number of running decentralized controllers in the industry. The main steps involved in employing decentralized controllers can be summarized as follows:

- Control objectives formulation and plant modeling.
- Control structure selection.
- Controller design.
- Simulation or pilot plant experiments and Implementation.

Nearly all the textbooks on multivariable control theory deal only with the control system analysis and design. The important concept of control structure selection which is a key prerequisite for a successful industrial control strategy is almost unnoticed. Structure selection involves the following two main steps:

- Inputs and outputs selection.
- Control configuration selection or the input-output pairing problem.

This book focuses on control configuration selection or the input-output pairing problem, which is defined as the procedure of selecting the appropriate input and output pair for the design of SISO (or block) controllers. An improper input-output pairing selection can be detrimental to closed loop stability and performance. The main objective in selecting an appropriate control configuration is to minimize the loop interactions. This will facilitate the single input single output independent design that is the basis of many industrial control systems.

This book reviews the main control configuration selection methods available in the literature. The seminal work of Bristol, the Relative Gain Array (RGA) introduced in 1966 is the first method proposed for input-output paring. This empirical method is the most widely control configuration selection strategy used in the practical designs of process control systems. During the nearly 4 decades after introducing the RGA, there are now many extensions and generalizations to the conventional RGA methodology.

Moreover, input-output pairing selection from other viewpoints such as the state space, passivity and soft computing has been introduced. Other approaches include, control configuration selection for nonlinear multivariable plants and pairing problem in the face of time varying parameters or plant uncertainties.

This book is the first monograph to deal in depth with the issue of control configuration selection. It is designed and written to serve the needs of a wide audience working in the area of decentralized control systems both in the academia and industry. The engineers and students interested in control configuration selection techniques, engineers already acquainted with the basic concepts and willing to be familiar with the more advanced strategies and postgraduate students who in addition to the aforementioned points want to get to the frontiers of research in the field may find this book useful.

The book achieves these objectives by reviewing and explaining the concepts involved in the techniques, starting with the RGA and to the latest methodologies available in the literature. Worked examples and simulation results are also available to further explain and clarify the main points. This book will also be useful to process control engineers, postgraduate students in process, electrical, and mechanical engineering disciplines. It is assumed that the reader is acquainted with a basic knowledge about linear and nonlinear system theory, analysis and design of linear multivariable control systems.

Most of the numerical examples are solved using MATLAB software and are available over the internet. Also, corrections and comments can be accessed from the authors' homepages:

- http://saba.kntu.ac.ir/eecd/khakisedigh
- http://webpages.iust.ac.ir/b_moaveni

Feedback from our readers will be appreciated and will help in improving the materials in this book. Please send any questions and comments you may have to authors email addresses: sedigh@kntu.ac.ir and b_moaveni@iust.ac.ir.

Finally, we would like to acknowledge the support of all the people who in some way helped us during the preparation of this book. In particular, we would like to thank our colleagues and graduate students at K. N. Toosi University of Technology, Shahid Rajaee Teacher Training University, and the staff of the Springer-Verlag for their support. The whole manuscript has been revised by our post graduate student Mr Nima Monshizadeh Naini and his technical comments and proof reading the draft chapters of the book is very much appreciated. Last, but not least, we would like to sincerely thank our families for their encouragement, support and patience during the holidays and evenings devoted to this book, the moments that truly belonged to them.

Tehran
July 2009

Ali Khaki-Sedigh
Bijan Moaveni

Contents

List of Abbreviations

ARMA	Auto Regressive Moving Average
BRG	Block Relative Gain
CSTR	Continuous Stirred Tank Reactor
DBRG	Dynamic Block Relative Gain
DIC	Decentralized Integral Controllability
Digraph	Direct Graph
DIOPM	Dynamic Input-Output Pairing Matrix
DMRS	Decentralized Model Reference Schemes
DRGA	Dynamic Relative Gain Array
ERGA	Effective Relative Gain Array
GBDD	Generalized Block Diagonal Dominant
GDRG	Generalized Dynamic Relative Gain
GNI	Generalized Niederlinski Index
HIIA	Hankel Interaction Index Array
HSV	Hankel Singular Value
IC	Integral Controllability
ICI	Integral Controllability with Integrity
IM	Interaction Measure
IMC	Internal Model Control
LFT	Linear Fractional Transform
MIMO	Multi-Input Multi-Output
NBRG	Nonsquare Block Relative Gain
NDBRG	Nonsquare Dynamic Block Relative Gain
NI	Niederlinski Index
NRGA	Nonlinear Relative Gain Array
NSRG	Nonsquare Relative Gain
NSRGA	Nonsquare Relative Gain Array
PM	Participation Matrix
PRG	Partial Relative Gain
PRGA	Performance Relative Gain Array
QBDD	Quasi Block Diagonal Dominance
RGA	Relative Gain Array
RIA	Relative Interaction Array
RLS	Recursive Least Square
SIMO	Single-Input Multi-Output
SISO	Single-Input Single-Output

SSE	Sum Squared Error
SSV	Structured Singular Value
STR	Self Tuning Regulator
SV	Singular Value
upper-LFT	Upper Linear Fractional Transform

Chapter 1
Introduction

Advancement of complex industrial processes in the last century and their continued development resulted in *complex multivariable* or *large-scale plants* that imposed high demands on control systems technology. Control systems community proposed two solution packages to tackle the stability and performance requirements of such plants. The first is to derive the multivariable plant by a *central controller*. The other approach is the *decentralized control* structure. Each of the proposed control methodologies has their advantages and deficiencies. However, experiences of the last decades indicate the increasing application and interest in the decentralized control structures.

Two key steps in a successful decentralized control system design are the *control configuration selection* and the *controller tunings*. The later is the subject of many control textbooks, technical reports and research monographs. The first step is the main concern of this book and in this chapter, a brief outline of the *control configuration selection* or the *input-output pairing problem* is given. These two terms are used interchangeably in this text to describe the same problem. In this chapter, decentralized control systems are briefly reviewed and the advantages and their design process are discussed. Then, an overview of the control configuration selection in the decentralized design process is given. A short survey on the fundamental results in the Relative Gain Array (RGA) and other advanced control configuration selection techniques is provided. Also, the open problems and future trends in the control configuration selection are summarized. Finally, the aim and scope of the book is given.

1.1 Decentralized Control Systems

During the past decades, in spite of the remarkably notable theoretical achievements in the analysis and design of centralized multivariable control systems, decentralized control is still widely used in many industrial processes. Centralized controllers have non-diagonal transfer function matrices, which can deal with highly interactive loops in the process. This is achieved by counteracting the interactions through the elaborate design of all the controller transfer function matrix elements. While, in decentralized control, independent feedback controllers are used to control a subset of the plant outputs with a subset of the plant inputs.

A. Khaki-Sedigh and B. Moaveni: Control Configuration Selection, LNCIS 391, pp. 1–12.
springerlink.com

There is no communication between these independent measurements and feedback loops. This leads to diagonal or block diagonal controller transfer function matrices.

It is a recognized fact that the complexity and computational load of control system analysis and design for multivariable plants grows faster than its dimensions (Siljak 1996). Hence, it would be beneficial to decompose a multivariable plant into subsystems and perform independent designs for each subsystem depending on their dynamical behavior and interactions from other subsystems. Additionally, in complex multivariable industrial plants, an accurate and detailed model of the plant and its interactions are rarely available. This necessitates serious robustness considerations. It is well known that decentralized controllers are inherently more robust in the presence of uncertainties, interconnections and controller failures (Siljak 1996).

The main reasons beyond the extensive use of decentralized controllers can be summarized as follows:

- Easy implementation.
- Efficient maintenance.
- Simple tuning procedures.
- Robust behavior in the face of fault and model uncertainties.

The decentralized control system design is in general divided into six steps (Van de Wal and De Jager 2001). These include:

- **Control objectives formulation.** The objectives are expressed in time and/or frequency domain specifications.
- **Plant modeling.** Model types and structures are selected and input-output or internal state space plant descriptions are derived using analytical or system identification based approaches.
- **Control structure selection.** Control structure selection involves two steps. The first step is the inputs and outputs selection. The second step is the *control configuration selection* or the *input-output pairing problem*.
- **Controller design.** This includes the selection of control system design strategy, which is chosen according to the multivariable plant characteristics and closed loop performance specifications. This is the subject most dealt with in the control system courses and textbooks.
- **Simulation or pilot plant experiments.** Priori to hardware implementations, the design is verified and tested via extensive simulation and/or pilot plant experiments.
- **Implementation.** Finally, hardware implementation of the sensor/actuator and control package is performed.

In control structure selection, the determination of the adequate number, place, and sensor/actuator type is the input-output selection process. The system outputs are the set of *measured variables* and the *controlled variables*. The system inputs are the set of *manipulated inputs* and the *exogenous inputs*, such as disturbances, sensor noise, and reference inputs or the set-points. In the control structure selection framework, and in this book, *outputs* are the measured variables and *inputs* are the manipulated inputs. The importance of this step lies in the fact that many plant properties such as zeros locations, hardware issues (cost and maintenance), reliability and control complexity depend on the input-output selection.

The second part of the third step, i.e. the control configuration selection or the input-output pairing problem is the focus of this book and is only relevant in the decentralized control designs. In the following, the formal definitions of the two steps in the control structure selection are provided. A definition for the input-out selection is given in (Van de Wal and De Jager 2001) as:

> "*Select suitable variables u to be manipulated by the controller (plant inputs) and suitable variables y to be supplied to the controller (plant outputs).*"

This topic is extensively reviewed by Van de Wal and De Jager (2001), and also tackled in (Skogestad and Postlethwaite 2005). The second stage is the control configuration selection or the input-output pairing problem. This is defined as:

> *Procedure of selecting the input and output pair used to design the SISO (or block) controllers, which finally results in a decentralized diagonal (or block diagonal) controller.*

The key step, which is particular to decentralized control, is the control configuration selection or the input-output paring problem. An improper input-output pairing can lead to closed loop instability or poor closed loop performance. The main objective in selecting an appropriate control configuration has been to minimize the loop interactions. This would result in a multivariable plant close to being diagonally dominant, and would therefore facilitate the single input single output independent designs required in a decentralized control scheme. In addition to interactions, the choice of pairing will influence the effect of disturbances and the closed loop stability in the face of loop failures in decentralized control system. Various input-output pairing methods are available in the literature. These methodologies are briefly reviewed in the next section.

1.2 Control Configuration Selection: An Overview

The seminal work of Bristol, the *Relative Gain Array* (RGA), introduced in 1966 is the first method proposed to input-output paring (Bristol 1966). This empirical method is the most widely used control configuration selection strategy. However, many extensions, generalizations and input-output pairing selections from other viewpoints are now available. During the nearly 4 decades after introducing the RGA method, several methods like, the *Dynamic Relative Gain Array* (DRGA),

Effective Relative Gain Array (ERGA), *Partial Relative Gain* (PRG), *Relative Interaction Array* (RIA), *Performance Relative Gain Array* (PRGA), *Block Relative Gain* (BRG), *Dynamic Block Relative Gain* (DBRG), RGA application in non-square plants and unstable and/or non-minimum phase plants, *integrity*, *condition number* and related ideas, and interaction associated analysis are developed. *Structured Singular Value* (SSV) is introduced to develop new interaction measures by some authors. Also, recently passivity concept is employed for control configuration selection.

RGA and the subsequent analysis and methodologies are transfer function based. A dynamic extension of the RGA based on the state space model is now available. State space approach to input-output pairing problem using the Cross-Gramian matrix and the balanced realization concepts, trace of multiplied controllability and observability Gramian matrices, and the Hankel-Norm of subsystems are also introduced. The *Participation Matrix* (PM), the *Hankel Interaction Index Array* (HIIA), and the *Dynamic Input-Output Pairing Matrix* (DIOPM) are defined in this respect.

Further theoretical results in the field of input-output pairing problem include: optimal pairing, input-output pairing of time varying multivariable plants, input-output pairing of uncertain multivariable plants, and input-output pairing of nonlinear multivariable plants.

The control configuration selection is briefly reviewed in (Kinnaert 1995) and (Van de Wal and De Jager 1995).

1.2.1 The Relative Gain Array (RGA)

Let the linear multivariable plant be described by $G(s)=\left[g_{ij}(s)\right]$ $(i,j=1,\ldots,m)$, where $g_{ij}(s)$ is the open loop gain from the jth input to the ith output. Also, consider $h_{ij}(s)$ as the gain of the ith output to the jth input, where all the loops except the ith output are under tight control. Then, λ_{ij} the $(i,j)^{th}$ element of the RGA is defined as $\lambda_{ij}(s)=g_{ij}(s)/h_{ij}(s)$. It is shown in (Bristol 1966) that $\Lambda(s)=[\lambda_{ij}(s)]=G(s)\cdot{*}G^{-T}(s)$, where "$\cdot *$" is the element by element product. The *tight control* condition of all loops except the ith loop, can be only valid in a specific frequency range. Therefore, the RGA is usually defined in terms of the steady state step response matrix $G(0)$. This is due to the presence of integral control in many processes and the difficulty of obtaining $G(j\omega)$ at all frequencies (Bristol 1966). The main properties of the RGA can be summarized as follows:

- The RGA can be employed as a two-way interaction measure.
- The RGA is input and output scaling independent, $\Lambda(D_O G D_I)=\Lambda(G)$ where D_I and D_O are diagonal matrices.

- Summation of the rows and columns of the RGA are equal to unity.
- Any permutation of rows and columns in the plant transfer function matrix G, leads to the same permutation in the RGA.
- If the transfer function matrix $G(s)$ is triangular or diagonal, the RGA will be the identity matrix.
- Relative perturbations in the elements of G and in its inverse are related by $\frac{d[G^{-1}]_{ji}}{[G^{-1}]_{ji}} = -\lambda_{ij} \frac{dg_{ij}}{g_{ij}}$ and the individual elements of the RGA can be computed as $\lambda_{ij} = (-1)^{i+j} \frac{g_{ij} \det(G^{ij})}{\det(G)}$, where $[.]_{ij}$ denotes the $(i,j)^{th}$ element of the matrix and G^{ij} denotes the matrix obtained after removing the ith row and the jth column.
- For stable multivariable plants with no zeros and poles at $s = 0$, a relation between the plant non-minimum phase characteristics and the sign of $\lim_{s \to \infty} \lambda_{ij}(s) = \lambda_{ij}(\infty)$ and $\lambda_{ij}(0)$ is developed.
- Large RGA elements indicate high closed loop sensitivity and are also equivalent to large condition number, which is an indication of control difficulty.
- Relative changes in the elements of the transfer function matrix and the corresponding RGA elements are related as: $\frac{d\lambda_{ij}}{\lambda_{ij}} = (1 - \lambda_{ij}) \frac{dg_{ij}}{g_{ij}}$.
- In the case of a multivariable plant with diagonal integral controllers and $\lambda_{ij} < 0$, the closed loop plant will have one of the following properties: It is unstable, the loop corresponding to $\lambda_{ij} < 0$ is unstable, or if the loop corresponding to $\lambda_{ij} < 0$ is opened, the closed loop plant becomes unstable.

Integrity. *Integrity* is a key property of a decentralized control system that ensures the closed-loop stability as subsystem controllers are brought in and out of service (Bristol 1966). Niederlinski (1971) introduced an interesting theorem to solve the integrity problem. It is shown that an interacting linear multivariable control system with all controllers of an integrating type, for example I, PI and PID, is unstable if $\mathrm{NI} = \frac{\det[G(0)]}{\prod_i g_{ii}(0)} < 0$, which is defined as the Niederlinski index. This leads to two necessary conditions to achieve integrity in decentralized control: Diagonal elements of the RGA corresponding to the appropriate pair and the corresponding NI, be positive.

Extension to Nonsquare Multivariable Plants. The original RGA is only applicable to square multivariable plants. However, many industrial plants are inherently nonsquare. By defining a nonsquare relative gain array (NSRGA), the RGA is extended to the nonsquare multivariable plants. NSRGA is defined as the ratio of the open-loop gain to the closed-loop gain, when all loops other than the considered loop are under tight control in the least square sense. Plants having more outputs than inputs are considered and the control objective is stated as a minimization problem of the sum of squares of steady state errors of the outputs with fewer inputs. Also, NSRGA can be used for the selection of a square subsystem in the squaring down process. In this approach, outputs with small row sums of the NSRGA are eliminated.

Extension to Unstable Multivariable Plants. The RGA is defined for stable multivariable plants only. However, the input-output pairing analysis is extended for the open loop unstable multivariable plants in the literature. Using the developed theories a pairing rule is introduced for decentralized control and integrity of the unstable multivariable plants.

Input-Output Pairing Procedure with RGA. To end this section, the crucial points derived from the above discussions to choose the input-output pair are summarized:

- Choose the input-output pair so that the RGA diagonal elements are close to one (Bristol 1966).
- RGA diagonal elements and the NI corresponding to the input-output pair must be positive.
- Large RGA elements are not appropriate for input-output pairing.

1.2.2 Advanced Techniques

The advanced control configuration techniques can be categorized in two general classes:

1. The RGA extensions and generalizations.
2. Control configuration techniques based on new viewpoints such as state space approach, uncertainty considerations, direct nonlinear approach and soft computing methods.

The main RGA extensions and generalizations are developed to cope with the shortcomings and deficiencies of the RGA. These can be summarized as follows:

- **Dynamic Relative Gain Array** (DRGA)**.** DRGA is defined to extend the RGA notion to non-zero frequencies. The proposed method can give the correct pairing results and measures the interaction more accurately in certain cases were the RGA does not respond correctly. However, the decisive feature of the RGA for control configuration selection, i.e. its independence of controller design and disturbances and minimal dependence on plant models are lost in the DRGA by including detailed controller design in some methods.

- **Performance Relative Gain Array** (PRGA). PRGA is developed to overcome the deficiency of the original RGA in coupling identification in certain multivariable plants, e.g. the triangular multivariable plants. PRGA is dependent on row and column permutations and output scaling but independent of input scaling.
- **Partial Relative Gain** (PRG). PRG is introduced to overcome the ambiguity and problems associated with the RGA analysis for plants larger than 2×2, and to assist in selecting block-decentralized structures. PRG can be used as a tool to compare the different control configurations with each other and to reject the infeasible control structures.
- **Effective Relative Gain Array** (ERGA). ERGA is introduced to achieve minimal interaction within the interested frequency range and considers the finite control bandwidth.
- **Relative Interaction Array** (RIA). RIA is derived using the loop decomposition approach to transform the multivariable plant into SISO loops with loop interaction structurally embedded. RIA provides information on closed-loop stability, integrity and robustness.
- **SSV Approach to Input-Output Pairing.** Arguing that most control configuration techniques are not mathematically rigorous, some authors have used a SSV framework to develop new interaction measures with sound theoretical basis. Singular Values or the principal gains of the open loop transfer function matrix were initially used to evaluate the multivariable stability margins, and later adopted for loop selection in the steady state and possible achievable performance with sensitivity analysis in multivariable plants. Also, SSV methodology is extended to the frequency domain such that the pairing problem can be carried out over a range of frequencies that are of practical significance for the plant. It also provides dynamic measures for interaction and sensitivity.
- **Block Pairing and Block Relative Gain** (BRG). BRG provides a block pairing of inputs and outputs configurations. In this generalized framework, different types of subsystems with given sets of outputs and inputs are available which yield various block decentralized controllers. The effectiveness of BRG and block decentralized control is shown in the case of highly interactive multivariable plants. In such multivariable plants, to avoid the complexities associated with full scale centralized control, block decentralized controls are employed with relatively easier design, tuning, implementation, and maintenance and furthermore the closed loop plant is more fault tolerant. However, the number of candidate blocks and hence possible designs increases with the plant dimensions. This necessitates the use of BRG to give an estimate of the closed loop plant behavior prior to the design. Dynamic Block Relative Gain (DBRG) is also introduced. DBRG is developed using the Decentralized Model Reference Schemes (DMRS). The closed loop performance is studied by defining the performance deterioration index, which comes from the difference

between the achievable performance when subsystem interactions are ignored and the actual performance in the presence of interactions. Finally, Nonsquare Block Relative Gain (NBRG) is developed to generalize the BRG concept to nonsquare multivariable plants.

The other advanced developments in the control configuration selection process are mainly summarized as follows:

- **Input-Output pairing based on the Passivity theory.** In a rather different but promising approach to the control configuration selection, the concept of *passivity* has been recently adopted. The properties and applications of passivity theory and passive systems are well known in the stability analysis and control systems design. However, the concept has been newly introduced in interaction analysis. Contrasting the conventional pairing methods that use the generalized diagonal dominance, the proposed method introduces a frequency dependent passivity index to pair the system inputs and outputs according to their so-called *degree of passivity*. It is claimed that this index provides the best achievable closed loop performance under decentralized control. Experimental nature and the simple open loop step response tests required for the implementation of the method are the main advantages of the proposed methodology.
- **State space based control configuration selection methodologies.** State space models are useful to get insight into the internal relationships in the plant. Also, the interaction index derived from the state space models is superior to the traditional RGA index performance when the multivariable interaction has a non-monotonic behaviour in frequency. A dynamic extension of the RGA based on the state space model is available and the technique is applied to processes whose time constants are substantially different. The singular perturbation technique is applied to a singularly perturbed linear time invariant plant in the state space formulation and the system is decomposed into *fast* and *slow* subsystems. These approximately model the fast and slow modes of the plant. Hence, the effects of the fast dynamics, often neglected in practice, are studied on the dynamic interaction measure. Also, balanced realization is used for input-output pairing. Where, the cross-Gramian matrix determinant of the balanced realized description shows the relationship between the different input-output pairs. Although it gives superior results to the RGA in some cases, its dependence on the balanced realization presentation of the plant and the fact that the determinant of the cross-Gramian matrix has no physical interpretations to relate the inputs to the outputs are its main drawbacks. In another development, an interaction measure based on a dynamic model of the process in the state space form is introduced. This measure is applicable to square stable multivariable plants and is formulated in continuous and discrete-time domains. Interaction is quantified as a function of chosen channel bandwidths and an index is derived from system Gramians for input-output pairing. The proposed index, is called the participation matrix (PM). PM can also be used in a broader aspect as a measure of the achievable performance of a given multivariable

controller in decentralized or more complex structures. PM allows the possibility of gradually increasing the complexity of control structure beyond a diagonal or decentralized control, to achieve a desirable performance without employing a full-scale centralized control. A similar approach that uses the Hankle Interaction Index Array (HIIA) is based on the Hankle norm of the SISO elementary subsystems built from the original multivariable process. HIIA can quantify the frequency dependent interactions, which in turn leads to a solution of the input-output pairing problem. The main advantage of the HIIA is its physical meaning to quantify the relation between plants inputs-outputs and its solutions to the scaling problem. Finally, the Dynamical Input-Output Pairing Matrix (DIOPM) is another defined interaction measure, which provides an appropriate input-output pair in the state space framework.

- **Direct approaches to input-output pairing of nonlinear multivariable plants.** The input-output pairing of nonlinear multivariable plants is mainly performed by the indirect approach, where the plant is linearized around its operating points and any of the approaches applicable for linear multivariable systems are used in the corresponding operating point. The general direct control configuration selection of nonlinear multivariable plants is still an open problem. In an early approach, a digraph using relative order of the nonlinear plant is proposed to show the relation between the inputs and outputs of the nonlinear plants, but this digraph cannot evaluate the interaction among the inputs-outputs channels of nonlinear multivariable plants. In another approach, classical RGA idea is extended to the nonlinear multivariable plants. Also, nonlinear-RGA (NRGA) is defined and is applied to the affine nonlinear multivariable plants. The dependence of pairing in the nonlinear multivariable plants on their operating point is clearly demonstrated.
- **Control configuration selection methodologies in the presence of time varying or uncertain plant parameters.** Early robustness considerations reported in the literature are limited to sensitivity analysis of the RGA elements to model error, where a parameter is subject to certain changes. It is argued that large RGA elements imply high sensitivity to plant gain uncertainties and a simple necessary condition for robust stability in the face of gain uncertainty is also derived. However, it should be noted that the derivations and consequences are limited to sensitivity analysis and not a full-scale parameter change or structured/unstructured type uncertainties. Hence, in the face of unknown or uncertain multivariable plants, the input-output configuration of the plant may endure fundamental changes, which will severely degrade the decentralized controller performance. The well-known input-output pairing techniques are unable to analyze the effect of uncertainty on the input-output pairing and only recently, pairing methods are proposed for uncertain multivariable plants. The adopted approaches can be categorized in the following classes: Statistical description of uncertainty bound for the RGA; Unstructured uncertainty modeling introduced to obtain a bound on the worst case magnitude of the relative gain; A graphical method to identify the change in the input-output pairs in the

presence of structured uncertainty; On-line identification and adaptive input-output pairing to encounter parameter changes. *Robust integrity* for uncertain multivariable plants is also defined and its relation to pairing and the RGA are explored.

- **Soft computing approaches.** *Soft Computing* methods have recently been employed to solve the input-output pairing problem. *Neural Networks*, *Genetic algorithms* and *Fuzzy logics* are the soft computing tools used in input-output pairing. Neural networks are used to measure the interaction and introduce the appropriate input-output pair to decentralized control. The neural approach uses the basic definitions of the RGA and extends the ideas for general time varying and nonlinear plants using neural networks. In this method, two neural network identifiers in forward and backward directions identify the plant and compute the Neural RGA. Genetic algorithms are used to introduce the optimal input-output pairing (French et al. 1995). On the other hand, fuzzy logic tackles the input-output pairing problem in terms of a linguistic fuzzy approach. Linguistic decoupling of complex multivariable plants is also introduced. Fuzzy modeling can be used to derive low-order Takagi-Sugeno (TS) type fuzzy models to describe complex nonlinear multivariable processes. The RGA and output sensitivity function methodologies are employed for the interaction analysis in multivariable TS fuzzy models. The novelty of the method lies in the fact that a single RGA cannot be used for the interaction analysis of TS type fuzzy models (Mollov et al. 2001). This is done by exploring the specific TS structure to get a series of RGAs for interaction analysis. When all model outputs have similar antecedent structure, RGAs can be computed for separate rules. Otherwise, RGA is calculated point-wise, combining the degrees of fulfillment for the point with the corresponding rule consequents. Fuzzy basis function network (FBFN) modeling based on a set of linguistic IF-THEN rules or data, is also introduced to model the complex nonlinear multivariable plants (Xu and Shin 2007). Then a steady state RGA is used to locally analyze the interactions of the plant around the operating points. Thus, the FBFN model is first linearized around the operating points and the RGA will be locally calculated based on the local linearization. This is performed for both square and nonsquare nonlinear multivariable plants.

1.3 Open Problems and Future Trends

This section provides a succinct look at the open problems and possible future research in the area of the input-output pairing problem. These are outlined below:

- Development of a unified framework for control structure selection, where the input-output selection and the control configuration selection or the input-output pairing process are treated in a fused theoretical approach.
- It is easily shown that the triangular transfer function matrix leads to the RGA as the unit matrix. However, the converse conjecture for an RGA close to the unit matrix is not rigorously verified.

- In multivariable plants with a number of possible control configurations, two approaches or a combination of them can be followed. First, is the *optimal* input-output pairing (or block pairing) selection. Second, is the *integration* of the control configuration selection with properties such as fixed modes, transmission zeros, integrity, robustness and interaction margins.
- The present results in both structured and unstructured uncertainties need more elaborations in several directions, including less conservative conditions for possible input-output pair change in the face of uncertainties, input-output pairing and robust integrity, consideration of multiplicative uncertainties and establishing necessary and sufficient conditions.
- Extension of block pairing results to the case of uncertain multivariable plants.
- Development of interaction measures and control configuration selection methods for nonsquare and/or unstable multivariable plants in a state space framework.
- Design of stable and robust adaptive pairing techniques to encounter pairing changes during the multivariable plant operation.
- Development of a rigorous treatment of input-output pairing in nonlinear multivariable plants, that directly provides the input-output pairs. An alternative possible approach could be the extension of passivity results to the nonlinear multivariable plants.
- Extension of the present results on affine nonlinear multivariable plants to the non-affine problems.
- Proposing convergent neural structures for further extending the idea of neural RGA and development of other soft computing type definitions for input-output pairing of nonlinear time varying multivariable plants.

1.4 Scope of the Book

This book provides an extensive review of the present available methodologies for the control configuration selection. It starts with an introduction in chapter 1 that briefly outlines the virtues of decentralized control. It then gives an overview of the control configuration problem from the orthodox RGA paradigm to the more advanced mathematically sound techniques. Open problems and future trends are outlined at the end of chapter 1. Chapter 2 is directed towards a uniform in depth treatment of the results available on the RGA theory. This includes the basic definitions, properties, pairing analysis and worked examples. The RGA extensions to nonsquare and unstable multivariable plants are also provided. Benchmark examples are adopted to show the main points raised in different sections. Chapter 3 is devoted to more advanced extensions of the RGA. It is shown that the RGA shortcomings can be resolved using the developed ideas. Structured singular value and passivity concepts are tackled in chapter 4. State spaces techniques have their own attractive features. The internal viewpoint that is provided, resolves many of the problems that may be associated with the more classically oriented pairing methods. After providing a short summary of the main necessary results from the linear systems theory, chapter 5 gives a detailed account of the present state space based pairing methodologies. Starting the pairing process directly with the nonlinear equations to describe the multivariable plant is an important development in control configuration selection. Chapter 6 deals with the problem of nonlinear multivariable plants. Comparison results with the linear analysis are also given. Uncertainty and robustness is a fundamental concept in any practical and serious control systems analysis. This issue has been the focal point of some researches in the pairing analysis during the past decades. However, only recently the structured and unstructured uncertainty have been completely treated. Chapter 7 starts with the RGA sensitivity analysis and then continues with both uncertainty types and considers their effects on the pairing analysis. Adaptive input-output pairing is another issue dealt with in chapter 7.

The book uses the MATLAB package for the examples in the text. The interested reader can find the M-files associated with the examples in http://saba.kntu.ac.ir/eecd/khakisedigh/books/controlconfiguration.htm. The examples are selected to present the main ideas and are mostly chosen form the cited literature as the benchmark problems in control configuration selection. The transfer function matrix and state space models of the plants used in the examples are gathered in the appendix for quick reference. Although the book does not describe in detail any specific controller design or tuning methodology, it uses closed loop responses to show the pitfalls or potency of a particular pairing scheme.

References

Bristol, E.H.: On a new measure of interaction for multivariable process control. IEEE Trans. Autom. Control. 11, 133–134 (1966)

French, I.G., Ho, C.K.S., Cox, C.S.: Genetic algorithms in controller structure selection. In: Proceeding of the Conference on Genetic Algorithms in Engineering Systems: Innovations and Applications, Sheffield, UK (1995)

Kinnaert, M.: Interaction measures and pairing of controlled and manipulated variables for multiple-input-multiple-output systems: a survey. Journal A 36, 15–23 (1995)

Mollov, S., Babuska, R., Verbruggen, H.B.: Analysis of interaction in MIMO Takagi-Sugeno fuzzy models. In: Proceeding of the 10th IEEE International Conference on Fuzzy Systems, Melbourne, Australia, vol. 3, pp. 769–773 (2001)

Siljak, D.D.: Decentralized control and computations: status and prospects. Annu. Rev. Control 20, 131–141 (1996)

Skogestad, S., Postlethwaite, I.: Multivariable feedback control analysis and design. Wiley, Chichester (2005)

Van de Wal, M., De Jager, B.: Control structure design: a survey. In: Proceeding of the American Control Conference, vol. 1, pp. 225–229 (1995)

Van de Wal, M., De Jager, B.: A review of methods for input/output selection. Automatica 37, 487–510 (2001)

Xu, C., Shin, Y.C.: Interaction analysis for MIMO nonlinear systems based on a fuzzy basis function network model. Fuzzy Set. Syst. 158, 2013–2025 (2007)

Chapter 2
Control Configuration Selection of Linear Multivariable Plants: The RGA Approach

2.1 Introduction

The RGA was proposed by Bristol in 1966 to facilitate the design of decentralized control systems by determining the control system configurations with minimal interactions.

The RGA or the *Holy Matrix* as named by Kariwala and Hovd (2006) is among the first systematic tools employed for interaction analysis and input-output pairing in linear multivariable plants. It is still the most widely used technique in the industry. The original version of the RGA only requires the steady state gain matrix of the plant, which can be easily determined by performing open loop step response tests on the complex multivariable plant. This property makes it attractive for practicing engineers, and is a key reason for its widespread use. This chapter provides an overview of the basic RGA concepts.

This chapter is organized as follows. In section 2.2, the RGA fundamentals including the basic definitions and derivations, the RGA properties, and finally the decisive issue of pairing and integrity are given. Section 2.3 provides the pairing rules derived from the RGA properties and shows the effectiveness of these rules in three worked examples. The RGA was originally developed for stable multivariable plants with equal number of inputs and outputs, i.e. square multivariable plants. The RGA extensions to unstable and nonsquare multivariable plants are given in section 2.4.

2.2 The RGA Fundamentals

This section presents a relatively detailed account of the RGA fundamentals. The RGA definition is derived from the first principles under the tight control assumption. Then, a list of the major RGA properties and their proofs are provided. The integrity issue, actuator and sensor failures are also discussed.

2.2.1 The Basic Principles

Consider the linear multivariable plant described by the following transfer function matrix model, obtained by linearizing the plant around an operating point

A. Khaki-Sedigh and B. Moaveni: Control Configuration Selection, LNCIS 391, pp. 13–55.
springerlink.com

$$G(s) = \left[g_{ij}(s) \right] \qquad i, j = 1, \ldots, m \tag{2.1}$$

where $g_{ij}(s)$ is the open loop gain from the jth input to i^{th} output. Let $Y(s) = \left[y_i(s) \right]$ $i = 1, \cdots, m$ be the output vector and $U(s) = [u_j(s)]$ $j = 1, \cdots, m$ be the input vector. In the decentralized control of multivariable plants, it is important to know that "how the transfer function between a specific output $y_i(s)$ and input $u_j(s)$ is affected by control of other variables?". The RGA measure proposed by Bristol answers this question. To derive the RGA matrix, we must first define the notion of *relative gain* as given by Bristol (1966). Assume that, $u_k = 0 \ \ \forall k \neq j$, i.e. the loops are open and the effect of u_j on the ith output y_i is considered. This gives the steady state gain between the jth input and the ith output in the open loop, denoted by $g_{ij}(0)$. On the other hand, in the case of closed loop regulation let $y_k = 0 \ \ \forall k \neq i$, i.e. keeping all the outputs constant except the ith output, and assuming *tight control* in all other channels. This steady state gain is denoted by $h_{ij}(0)$ and is the gain of the ith output to the jth input, where all the loops except the ith output are under tight control. In the following derivations, only steady state values are considered and these will be denoted by the same characters without the s that is $G(0) = G$, and $g_{ij}(0) = g_{ij}$. Then, the relative gain, which is a dimensionless number and denoted by λ_{ij} is defined as follows

$$\lambda_{ij} = g_{ij} / h_{ij} \tag{2.2}$$

and the RGA is defined as

$$\Lambda = \left[\lambda_{ij} \right] \qquad i, j = 1, \cdots, m \tag{2.3}$$

To obtain a straightforward formulae for computing the RGA, note that $Y = GU$ and $U = G^{-1}Y$. Let, $G^{-1} = \hat{G} = [\hat{g}_{ij}]$. Then, a change in the jth input will lead to a change in all the outputs. That is

$$\delta u_j = \sum_{k=1}^{m} \hat{g}_{jk} \delta y_k \tag{2.4}$$

However, if the condition of tight control is satisfied, we have

$$\delta u_j = \hat{g}_{ji} \delta y_i \tag{2.5}$$

This gives

$$\delta y_i = \frac{1}{\hat{g}_{ji}} \delta u_j \tag{2.6}$$

Hence

$$h_{ij} = \frac{1}{\hat{g}_{ji}} \tag{2.7}$$

Therefore, equations (2.2) and (2.7) give

$$\lambda_{ij} = g_{ij} \hat{g}_{ji} \tag{2.8}$$

and equation (2.3) can thus be rewritten as

$$\Lambda(G) = [\lambda_{ij}] = G \cdot\!* G^{-T} \tag{2.9}$$

where "$\cdot *$" is the element by element product. This is the steady state RGA and the same idea is applied to other frequencies. The RGA is normalized, that is the sum of the elements of each row and column adds up to one.

Relative gain is by definition an indication of *interaction* in the multivariable plants, i.e. the perturbations caused by the closed control loops. It is clear that the pairs with relative gains equal to one will have the least interaction from the other loops. Hence, the farther the relative gain from unity, the stronger the interaction.

Example 2.2.1

Consider the following distillation column transfer function matrix (Wood and Berry 1973)

$$\begin{bmatrix} X_D(s) \\ X_B(s) \end{bmatrix} = \underbrace{\begin{bmatrix} \frac{12.8e^{-s}}{16.7s+1} & \frac{-18.9e^{-s}}{21s+1} \\ \frac{6.6e^{-7s}}{10.9s+1} & \frac{-19.4e^{-s}}{14.4s+1} \end{bmatrix}}_{G(s)} \begin{bmatrix} R(s) \\ S(s) \end{bmatrix} \tag{2.10}$$

the steady state matrix G is

$$G = \begin{bmatrix} g_{11} & g_{12} \\ g_{21} & g_{22} \end{bmatrix} = \begin{bmatrix} 12.8 & -18.9 \\ 6.6 & -19.4 \end{bmatrix} \tag{2.11}$$

which gives

$$\hat{G} = \begin{bmatrix} \hat{g}_{11} & \hat{g}_{12} \\ \hat{g}_{21} & \hat{g}_{22} \end{bmatrix} = \begin{bmatrix} 0.1570 & -0.1529 \\ 0.0534 & -0.1036 \end{bmatrix} \tag{2.12}$$

so, the RGA is obtained as

$$\Lambda(G) = \begin{bmatrix} g_{11}\hat{g}_{11} & g_{12}\hat{g}_{21} \\ g_{21}\hat{g}_{12} & g_{22}\hat{g}_{22} \end{bmatrix} = \begin{bmatrix} 2.0094 & -1.0094 \\ -1.0094 & 2.0094 \end{bmatrix} \tag{2.13}$$

■

Example 2.2.2

Consider a 2×2 multivariable plant described by the model

$$\begin{aligned} y_1(s) &= g_{11}(s)u_1(s) + g_{12}(s)u_2(s) \\ y_2(s) &= g_{21}(s)u_1(s) + g_{22}(s)u_2(s) \end{aligned} \tag{2.14}$$

The transfer function between $y_1(s)$ and $u_1(s)$ with no control on $y_2(s)$, i.e. $u_2(s) = 0$ is $g_{11}(s)$. Also, when $y_2(s)$ is under tight control, i.e. $y_2(s) = 0$, we have by eliminating $u_2(s)$

$$y_1(s) = g_{11}(s)\left[1 - \frac{g_{12}(s)g_{21}(s)}{g_{11}(s)g_{22}(s)}\right]u_1(s) = h_{11}(s)u_1(s) \tag{2.15}$$

where h_{11} is as defined in equation (2.7). Thus, the RGA can be written in the following form

$$\Lambda = \begin{bmatrix} \lambda & 1-\lambda \\ 1-\lambda & \lambda \end{bmatrix} \tag{2.16}$$

where

$$\lambda = \frac{1}{1 - \dfrac{g_{12}g_{21}}{g_{11}g_{22}}} \in \Re \tag{2.17}$$

Properties of the 2×2 multivariable plants can be studied for different values of λ:

1. Let $\lambda = 1$, then open loop and closed loop gains are the same and the interaction does not affect the open loop gain between u_1 and y_1. In this case the loops are *decoupled*. Therefore, the appropriate pairing is $(u_1 - y_1, u_2 - y_2)$.

2. Let $\lambda = 0$, it is obvious that $g_{11} = 0$ and the first input has no effect on the first output. It can only control the second output. Therefore, the appropriate pairing is $(u_1 - y_2, u_2 - y_1)$.
3. Let $0 < \lambda < 1$, it is obvious from equation (2.15) that closing the loop $(u_2 - y_2)$ increases the gain between u_1 and y_1, depending on the value of λ, the size of interaction changes. The worst interaction case occurs when $\lambda = 0.5$.
4. Let $\lambda > 1$, the interaction is significant and the condition worsens for higher values of λ, i.e. closing the loop $(u_2 - y_2)$ decreases the gain between u_1 and y_1.
5. Let $\lambda < 0$, the gain between u_1 and y_1 changes sign when the loop $(u_2 - y_2)$ is closed. This causes serious control problems. Hence, $1 - \lambda > 0$ and $(u_1 - y_2, u_2 - y_1)$ is the recommended pairing.

■

Example 2.2.3

Consider the following 2×2 multivariable plant

$$Y(s) = \begin{bmatrix} \dfrac{1}{0.1s+1} & \dfrac{1}{s+1} \\ \dfrac{2}{s+1} & \dfrac{1}{0.5s+1} \end{bmatrix} U(s)$$

A step change in u_1 with $u_2 = 0$, gives $y_1(0) = 1$, hence $\dfrac{\delta y_1(0)}{\delta u_1(0)} = g_{11}(0) = 1$.

Now, to keep y_2 constant, i.e. $y_2 = 0$ under tight control by u_2, let

$$u_2(s) = -2\frac{0.5s+1}{s+1}u_1(s)$$

which gives $y_2 = 0$. Therefore, substituting this into the plant equations gives

$$y_1(s) = \left(\frac{1}{0.1s+1} + \frac{1}{s+1} \times \frac{-2(0.5s+1)}{s+1}\right)u_1(s)$$

For a unit step change in u_1, the steady state output $y_1(0)$ is calculated from the final value theorem, which is $y_1(0) = -1$. Hence, $\left.\dfrac{\delta y_1(0)}{\delta u_1(0)}\right|_{y_2=0} = h_{11}(0) = -1$

and therefore $\lambda_{11} = g_{11}/h_{11} = -1$. Thus, the RGA is

$$\Lambda = \begin{bmatrix} -1 & 2 \\ 2 & -1 \end{bmatrix}$$

This gives $(u_1 - y_2, u_2 - y_1)$ as the recommended pair.

■

Similarly, the frequency dependent RGA for the linear multivariable plant given by equation (2.1) is (Bristol 1966)

$$\Lambda(s) = [\lambda_{ij}(s)] = G(s) \cdot * G^{-T}(s) \tag{2.18}$$

The 'tight control' condition of all loops except the ith loop, can be only valid in a specific frequency range. Therefore, the RGA is usually defined in terms of the steady state step response matrix $G(0)$. This is due to the presence of integral control in many processes and the difficulty of obtaining $G(j\omega)$ at all frequencies (Bristol 1966). However, the RGA at specific frequencies close to the cross-over frequencies can also be valuable. The RGA fundamentals can also be found in (Maciejowski 1989), (Skogestad and Postlethwaite 2005) and (Kinnaert 1995).

2.2.2 The RGA Properties

In this section, the main RGA properties are summarized. These properties can facilitate the RGA computation and input-output pairing.

Property 1. We have, $GG^{-1} = I$ which gives $\sum_{j=1}^{m} g_{ij}\hat{g}_{ji} = 1$. Using equation (2.8) $\sum_{j=1}^{m} \lambda_{ij} = 1$ and similarly, $\sum_{i=1}^{m} \lambda_{ij} = 1$. This leads to the first RGA property,

Summation of the elements in each row or column of the RGA is unity.

A direct consequence of this property is,

At least one eigenvalue and one singular value of the RGA is equal to 1.

Property 2. Permutation of the rows and columns in the transfer function matrix leads to similar permutations in the RGA. The permutation matrix P is an elementary matrix which is found by performing the elementary operations of interchanging any two rows or columns on the $m \times m$ unit matrix. The row and column permutations in G can therefore be represented by the elementary matrix multiplications P_1GP_2, and $\Lambda(P_1GP_2) = P_1\Lambda(G)P_2$. Hence, the second RGA property is derived,

Any permutation of rows and columns in the plant transfer function matrix results in the same permutation in the RGA.

Property 3. Scaling is an important issue in practical control systems design. The scaled transfer function matrix can be represented by $G_s = D_1GD_2$, where $D_1 = diag\{d_{1i}\}$ and $D_2 = diag\{d_{2i}\}$ $(i = 1, \cdots, m)$ are nonsingular diagonal matrices. Then, the elements of G_s and $\hat{G}_s$ are $g_{sij} = d_{1i}g_{ij}d_{2j}$ and

$\hat{g}_{sij} = \frac{1}{d_{2j}} \hat{g}_{ij} \frac{1}{d_{1i}}$, respectively. Hence, $\Lambda(G_s) = \Lambda(G)$ which is the third RGA property,

The RGA is input and output scaling independent.

Note. Input or output delays can be represented by diagonal scaling matrices, which do not affect the RGA elements. For example, adding a time delay to each output can be represented by the scaling matrix $D_1 = diag(e^{-T_i s})$, and we have $\Lambda(D_1 G) = \Lambda(G)$.

Property 4. The RGA of diagonal transfer function matrices, i.e. completely decoupled plants, is trivially the identity matrix. In the case of upper triangular transfer function matrices, we have $g_{ij} = 0$ for $i > j$. This is known as one way interaction. Note that $\hat{G} = \frac{1}{\det(G)} \text{adj}(G)$ and $\text{adj}(G) = \left[c_{ij}\right]^T$, where $c_{ij} = (-1)^{i+j} \det G^{ij}$ are the cofactors of G and G^{ij} is obtained by removing the ith row and j^{th} column of G. Then, it is obvious that $c_{ij} = 0$ for $j > i$ and all the off diagonal elements of the RGA are zero. We can therefore conclude that $\Lambda(G) = I$ for triangular multivariable plants. Also, it can be shown that in the case of 2×2 and 3×3 multivariable plants, $\Lambda(G) = I$ implies that the plant is triangular for 2×2 and triangular or *essentially* triangular for 3×3 multivariable plants. Essentially triangular multivariable plants are defined as the multivariable plants that can be rendered triangular by column or row permutations. This is wrongly generalized to general $m \times m$ multivariable plants by many researchers (Hovd and Skogestad 1992) and (Skogestad and Postlethwaite 2005). This issue is dealt with in (Kariwala and Hovd 2006) and the following counterexample is provided by the authors from an earlier work on the subject.

Example 2.2.4

Consider the following matrix

$$G = \begin{bmatrix} 1 & 1 & 0 & 0 \\ 0 & 1 & 1 & 1 \\ 1 & 1 & \varepsilon & 0 \\ 0 & 0 & 1 & 1 \end{bmatrix}$$

It is easily shown that $\Lambda(G) = I$ for all real non zero ε, while the matrix itself is neither triangular nor essentially triangular.

■

Hence, the fourth RGA property can be stated as follows,

In the case of 2×2 *multivariable plants,* $\Lambda(G)=I$ *if and only if the plant is triangular. In the case of* 3×3 *multivariable plants, if* $\Lambda(G)=I$ *then the plant is triangular or essentially triangular. However, triangular multivariable plants of any order will have an identity matrix RGA.*

Property 5. The fifth algebraic RGA property is, The individual elements of the RGA, can be computed as

$$\lambda_{ij} = (-1)^{i+j} \frac{g_{ij} \det(G^{ij})}{\det(G)}.$$

This is easily shown as

$$\lambda_{ij} = g_{ij} \hat{g}_{ji} = g_{ij} \frac{c_{ij}}{\det(G)} = g_{ij} (-1)^{i+j} \frac{\det(G^{ij})}{\det(G)} \tag{2.19}$$

where c_{ij} denotes the cofactors of G.

Property 6. By defining G^{i*} as the ith row of G less g_{ij}, and defining G^{*j} as the jth column of G less g_{ij}, we have

$$\hat{g}_{ji} = \frac{1}{g_{ij} - G^{i*}(G^{ij})^{-1}G^{*j}}$$

which gives

$$\lambda_{ij} = g_{ij}\hat{g}_{ji} = \frac{g_{ij}}{g_{ij} - G^{i*}(G^{ij})^{-1}G^{*j}} \tag{2.20}$$

hence

$$\frac{d\hat{g}_{ji}}{dg_{ij}} = \frac{-1}{(g_{ij} - G^{i*}(G^{ij})^{-1}G^{*j})^2} = -\hat{g}_{ji} \frac{\lambda_{ij}}{g_{ij}}$$

The sixth algebraic RGA property now follows (Grosdidier et al. 1985),

Relative perturbations in the elements of G and *in its inverse are related by*

$$\frac{d\hat{g}_{ji}}{\hat{g}_{ji}} = -\lambda_{ij} \frac{dg_{ij}}{g_{ij}} \tag{2.21}$$

It follows from this property that if the $(i, j)^{th}$ element of the RGA is large, a change in g_{ij} will cause larger relative changes in $\hat{g}_{ji}$, which is undesirable.

Property 7. To show the next property, the expansion of $\det(G)$ can be written as

$$\det(G) = \sum_{i=1}^{m} (-1)^{i+j} g_{ij} \det(G^{ij}) \tag{2.22}$$

Consider a perturbation in the $(i, j)^{th}$ element of G and denote the perturbed plant transfer function matrix as $\tilde{G}$. Then

$$\det(\tilde{G}) = \sum_{i=1}^{m} (-1)^{i+j} \tilde{g}_{ij} \det(G^{ij}) \tag{2.23}$$

Equations (2.22) and (2.23) give

$$\det(\tilde{G}) - \det(G) = (-1)^{i+j} (\tilde{g}_{ij} - g_{ij}) \det(G^{ij}) \tag{2.24}$$

If the perturbation is

$$\tilde{g}_{ij} = g_{ij} (1 - \frac{1}{\lambda_{ij}}) \tag{2.25}$$

Then, using equations (2.19) and (2.25) in equation (2.24) gives

$$\det(\tilde{G}) = \det(G) - (-1)^{i+j} \frac{g_{ij}}{\lambda_{ij}} \det(G^{ij}) = 0 \tag{2.26}$$

This implies that large RGA elements are indicative of sensitivity to small relative errors in the individual elements of the transfer function matrix. The seventh RGA property is thus stated as follows,

If a single element of the transfer function matrix $G = [g_{ij}]$ is changed to $g_{ij} (1 - \frac{1}{\lambda_{ij}})$, then the perturbed transfer function matrix becomes singular.

This property has several important control implications. These can be stated as (Hovd and Skogestad 1992):

- If the relative uncertainty in an element at a given frequency is larger than $\frac{1}{\lambda_{ij}(j\omega)}$, then the plant may have zeros on the imaginary axis and unstable zeros at this frequency. However, in real processes element by element uncertainty rarely occurs, due to the coupling in the plant.

- In many industrial applications, system identification techniques are used to model the individual elements of the transfer function matrix. If large RGA elements are observed in the bandwidth where the model is to be used, incorrect results may be deducted, and other plant information must be considered to interpret the results.

Property 8. Diagonal input uncertainty perturbs the nominal transfer function matrix $G(s)$ to $\tilde{G}(s)$ where (Hovd and Skogestad 1992)

$$\tilde{G}(s) = G(s)(I + \Delta) \tag{2.27}$$

and $\Delta = diag(\Delta_i)$ represents the relative uncertainties in different input channels. For an inverse-based decoupling controller as

$$C(s) = G^{-1}(s)K(s), \qquad K(s) = diag\{k_i(s)\} \tag{2.28}$$

The compensated plant is

$$\tilde{G}(s)C(s) = (I + G(s)\Delta G^{-1}(s))K(s) \tag{2.29}$$

We have

$$(G(s)\Delta G^{-1}(s))_{ii} = \sum_{j=1}^{m} \lambda_{ij}(G)\Delta_j \tag{2.30}$$

which results in the eighth RGA property as,

Severe sensitivity results from large RGA elements with inverse based controllers.

Property 9. Differentiating equation (2.20) gives (Grosdidier et al. 1985), *Relative changes in the elements of the transfer function matrix and the corresponding RGA elements are related as:* $\frac{d\lambda_{ij}}{\lambda_{ij}} = (1 - \lambda_{ij})\frac{dg_{ij}}{g_{ij}}$.

Property 10. In (Bristol 1966), it is claimed that "the transfer function between the jth input and the ith output with all other loops closed will be non-minimum phase or unstable if λ_{ij} is negative". Later it was shown by a counterexample in (Grosdidier et al. 1985) that this is not in general true. This counterexample is given below.

Example 2.2.5

Consider the following transfer function matrices,

$$G_1(s) = \frac{1}{s+1}\begin{bmatrix} \pm s + 2 & 1 \\ 3 & 1 \end{bmatrix}$$

It can be shown that the corresponding first RGA elements are, $\lambda_{11}(s)\Big|_{g_{11}(s)=\frac{s+2}{s+1}} = \frac{s+2}{s-1}$ and $\lambda_{11}(s)\Big|_{g_{11}(s)=\frac{-s+2}{s+1}} = -\frac{-s+2}{s+1}$. Hence, $\lambda_{11}(0) = -2$ in both cases. Note that the zero polynomials of the transfer functions between the first input and the first output with the second loop closed are $\pm s-1$, that is non-minimum phase for $g_{11}(s) = \frac{s+2}{s+1}$ and minimum phase for $g_{11}(s) = \frac{-s+2}{s+1}$. Also, the multivariable plants have a transmission zero at $s = 1$ and $s = -1$, respectively.

Next, consider the following transfer function matrices,

$$G_2(s) = \frac{1}{s+1}\begin{bmatrix} \pm s+2 & 1 \\ -3 & 1 \end{bmatrix}$$

It can similarly be shown that $\lambda_{11}(0) = 0.4$, in both cases and there will be no sign change. However, the multivariable plants have a transmission zero at $s = 5$ and $s = -5$, respectively.

■

This example clearly indicates that positive or negative signs of $\lambda_{11}(0)$ can correspond to stable or unstable transmission zeros. However, it can be shown that there is indeed a relationship between non-minimum phase zeros and the sign change of the respective RGA elements at zero and infinite frequencies (Skogestad and Hovd 1990) and (Hovd and Skogestad 1992). Let $s = j\omega$ in property 5,

$$\lambda_{ij}(j\omega) = (-1)^{i+j} \frac{g_{ij}(j\omega)\det(G^{ij}(j\omega))}{\det(G(j\omega))} \tag{2.31}$$

If $\lim_{s\to\infty} \lambda_{ij}(s)$ is finite and non zero, then numerator and denominator of equation (2.31) would be of the same degree. Then, a sign change of $\lambda_{ij}(\infty)$ and $\lambda_{ij}(0)$ would imply RHP-poles or zeros in $g_{ij}(j\omega)$, $\det(G^{ij}(j\omega))$ or $\det(G(j\omega))$. Note that for stable multivariable plants RHP-zeros of $\det(G(s))$ show the RHP transmission zeros of $G(s)$. However, if the plant is assumed stable with no zeros or poles at the origin, the following property can be deduced.

Consider a stable multivariable plant with the transfer function matrix $G(s)$ having no zeros and poles at $s = 0$. If $\lim_{s\to\infty} \lambda_{ij}(s) = \lambda_{ij}(\infty)$ and $\lambda_{ij}(0)$ have different sign then one of the following will be true: g_{ij} has a RHP zero, $G(s)$ has a RHP transmission zero or $G^{ij}(s)$ has a RHP transmission zero.

Note that any of the above conditions causes serious problems in the controlled plant. However, as noted in (Hovd and Skogestad 1992):

- This is only a sufficient condition and it is not a necessary condition.

Example 2.2.6

Consider the following stable transfer function matrix (Skogestad and Hovd 1990)

$$G(s)=\frac{1}{\tau s+1}\begin{bmatrix} s+1 & s+4 \\ 1 & 2 \end{bmatrix} \tag{2.32}$$

Then

$$\lambda_{11}(s)=\frac{2(s+1)}{s-2} \tag{2.33}$$

Hence, $\lambda_{11}(\infty)=2$ and $\lambda_{11}(0)=-1$. There is a sign change, and none of the diagonal elements have RHP-zeros. Hence, $G(s)$ must be non-minimum phase. Calculating the Smith Macmillan form of $G(s)$ shows a transmission zero at $s=2$.

■

2.2.3 *Pairing and Integrity*

Interaction in multivariable plants can deteriorate the closed loop properties by:

- Introducing disturbances between different loops.
- Obliterating the overall closed loop stability, even though the independently designed controllers are individually stable with desirable performance.

The RGA can be used as an interaction measure and the properties developed in section 2.2.2 will be used for this purpose. In this section, the role of the RGA in closed loop stability analysis will be studied. This can be of great importance in industrial applications because of the consequences of stability information in control system design and the simplicity of the RGA calculations. In the derivation of the RGA in section 2.2.1, the "tight or perfect control" assumption plays a key role. This is achieved in most practical control system design by *integral action*. This issue is dealt with in (Niederlinski 1971), (Grosdidier et al. 1985), (Chiu and Arkun 1990) and (Skogestad and Postlethwaite 2005). Hence, we begin our study with the plant under closed loop integral control. This guarantees the tight control assumption at steady state. The general multivariable integral control strategy is shown in figure 2.1.

Note that any feedback multivariable controller with built in integral action can be represented as shown in figure 2.1. In fact, the controller matrix is decomposed into an integrator matrix $\frac{k}{s}I$, where k is the positive tuning parameter, and a compensator matrix $C(s)$. This facilitates the generality of the stability analysis in

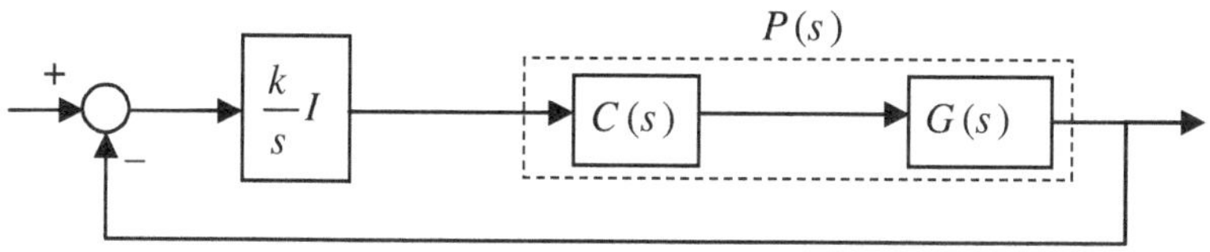

Fig. 2.1 Multivariable integral control.

the presence of integral action regardless of the employed particular compensation techniques. Let the open loop compensated plant matrix be denoted by $P(s)=G(s)C(s)$. However, $C(s)$ is designed such that $P(s)$ is a *proper* transfer function matrix, i.e. $\lim_{s\to\infty} P(s)$ is a constant matrix. It would be *strictly proper* if the resulting constant matrix is zero. Otherwise, it will be *improper*.

Definition 2.1. The plant $P(s)=G(s)C(s)$ is called *integral stabilizable* if there exists a $k>0$ such that the closed loop plant shown in figure 2.1 is stable and has zero steady state error for all constant inputs.

Zero steady state error condition discards the cancellation of the integrators by zeros. To investigate the closed loop stability of the multivariable integral control shown in figure 2.1, we first derive its characteristic equation. The closed loop characteristic equation is

$$\phi(s)\det(I+P(s)\frac{k}{s}I)=0 \tag{2.34}$$

where $\phi(s)$ is the characteristic equation of the loop gain $P(s)\frac{k}{s}I$. The transfer function matrix $P(s)$ can be rewritten as

$$P(s)=\frac{1}{d(s)}N(s) \tag{2.35}$$

where $N(s)$ is a polynomial matrix and $d(s)$ is the common denominator of the elements of $P(s)$. Substitution of equation (2.35) in (2.34) gives

$$\phi(s)\det(I+N(s)\frac{k}{sd(s)}I)=0$$

and

$$\frac{\phi(s)}{sd(s)}\det(sd(s)I+kN(s))=0 \tag{2.36}$$

Determinant expansion gives

$$\frac{\phi(s)}{sd(s)}(s^n d^n(s)+\cdots+k^n \det(N(0)))=0 \tag{2.37}$$

Assuming open loop stability the coefficient of the highest power of $d(s)$ is positive, and assuming that $P(s)$ is proper, the coefficient of the highest power of s in equation (2.37) will be the coefficient of the highest power of s in $d(s)$. The closed loop plant will be stable *only if* all the coefficients in equation (2.37) are positive. That is, only if $\det(N(0))$ and hence $\det(P(0))$ is positive. This is summarized as the following theorem (Grosdidier et al. 1985).

Theorem 2.1. Let $P(s)$ be a stable, proper and rational transfer function matrix. It is integral stabilizable only if $\det(P(0))>0$.

■

Example 2.2.7

Consider the multivariable plant

$$P(s)=\frac{1}{s+1}\begin{bmatrix}\pm 2 & 1\\ 1 & \pm 2\end{bmatrix}$$

which in both cases, $\det(P(0))=3$. The closed loop characteristic equation for the plant shown in figure 2.1 with $+2$ is

$$s^4+2s^3+(1+4k)s^2+4ks+3k^2=0$$

where it is stable for $k>0$. While, the closed loop characteristic equation of the closed loop plant for -2 is

$$s^4+2s^3+(1-4k)s^2-4ks+3k^2=0$$

and the corresponding closed loop plant is unstable for $k>0$.

■

This example indicates that Theorem 2.1 is only a necessary condition for integral stabilizability of multivariable plants. Although, it is a necessary and sufficient condition for SISO plants.

An important and desirable feature of a decentralized control system is *integrity*. Integrity is the property that ensures the closed-loop stability as subsystem controllers are brought in and out of service (Bristol 1966). Niederlinski (1971) introduced an interesting theorem to solve the integrity problem. However, it is shown in (Grosdidier et al. 1985) that the theorem in (Niederlinski 1971) is not stated correctly. The theorem presented in (Grosdidier et al. 1985) is as follows.

Theorem 2.2
Consider the multivariable closed loop plant shown in figure 2.1 with $C(s)$ diagonal. Also, assume that:

- $G(s)$ is stable.
- $P(s)=G(s)C(s)$ is a rational and proper matrix.
- All single input single output closed loops, obtained from opening any $m-1$ feedback loops, are stable.

Then, the system is unstable for all $k>0$ if

$$\frac{\det[G(0)]}{\prod_i g_{ii}(0)}<0 \tag{2.38}$$

Proof. Let $C(s)=diag\{c_1(s),\cdots,c_m(s)\}$, the loop gain for each individual loop is $c_i(s)g_{ii}(s)$ for $i=1,\cdots,m$. It is well known that each loop gain as a SISO plant can be stabilized with steady state tracking error if and only if $c_i(0)g_{ii}(0)>0$ for all i. It is obvious that positive feedback makes the closed loop unstable. Hence

$$\prod_{i=1}^{m} c_i(0)g_{ii}(0)>0 \tag{2.39}$$

On the other hand, theorem 2.1 states that $P(s)$ is not integral stabilizable if

$$\det(P(0))=\prod_{i=1}^{m} c_i(0)\det(G(0))<0 \tag{2.40}$$

Equations (2.39) and (2.40) give

$$\frac{\det(G(0))}{\prod_{i=1}^{m} g_{ii}(0)}<0 \tag{2.41}$$

■

Note 1. Equation (2.41) is a sufficient but not necessary condition. However, for SISO and 2×2 plants this is a necessary and sufficient condition. A proof for 2×2 plants can be found in (Grosdidier et al. 1985). The left hand side in equation (2.41) is called the *Niederlinski index* (NI) and is represented by NI[$G(0)$].

Note 2. We have

$$\begin{aligned}\mathrm{NI}[G(0)] &= \frac{\det(G(0))}{\prod_{i=1}^{m} g_{ii}(0)} = \frac{\det(G(0))}{\det(G^{ii}(0))g_{ii}(0)} \frac{\det(G^{ii}(0))}{\prod_{\substack{j=1\\ j\neq i}}^{m} g_{jj}(0)} \\ &= \frac{\det(G(0))}{\det(G^{ii}(0))g_{ii}(0)} \mathrm{NI}[G^{ii}(0)]\end{aligned} \tag{2.42}$$

Using equation (2.31), equation (2.42) is rewritten as

$$\mathrm{NI}[G(0)] = (\lambda_{ii}(0))^{-1}\mathrm{NI}[G^{ii}(0)] \tag{2.43}$$

Equation (2.43) is an important relationship between NI and the RGA (Chiu and Arkun 1990).

Example 2.2.8

Consider the following transfer function matrix

$$G(s) = \begin{bmatrix} 1 & 0 \\ 1 & \dfrac{-(s+1)}{2s+1} \end{bmatrix}$$

with a PI controller $\frac{k(s+1)}{s}I$. The closed loop characteristic equation is

$$[(1+k)s+k][(2-k)s^2+(1-2k)s-k]=0$$

And the closed loop plant is stable for $k>2$. However, $\det(P(0))=-1<0$. This points out the importance of the second condition in theorem 2.2, i.e. the *properness* of the compensated transfer function. In this case

$$P(s) = \begin{bmatrix} s+1 & 0 \\ s+1 & -\dfrac{(s+1)^2}{2s+1} \end{bmatrix}$$

which is an improper transfer function matrix. This critical point was missed out in the original work of (Niederlinski 1971).

■

The following example is adopted from (Chiu and Arkun 1990), which is originally from an earlier paper in this field.

Example 2.2.9

Consider the following steady state transfer function matrix

$$G(0)=\begin{bmatrix} -11.3 & -2.368 & -9.811 & 0.374 \\ 5.24 & 0.422 & 5.984 & -1.986 \\ -0.33 & 0.513 & 2.38 & 0.0204 \\ 4.48 & 15.54 & -11.3 & -0.176 \end{bmatrix}$$

and the corresponding RGA as

$$\Lambda=\begin{bmatrix} 1.0063 & -0.0314 & 0.1264 & -0.1013 \\ -0.1045 & 0.0003 & 0.0107 & 1.0935 \\ 0.1081 & 0.1630 & 0.7264 & 0.0025 \\ -0.0099 & 0.8680 & 0.1366 & 0.0054 \end{bmatrix}$$

diagonal elements of the RGA are $\{1.006 \quad 0.0003 \quad 0.726 \quad 0.005\}$ and the NI is -490.891. Then, neither of the original 4×4 or the reduced 3×3 plants can be stabilized. However, it is worth mentioning that by permuting the second and fourth rows, the NI of the permuted transfer function matrix is 1.1814, and the respective RGA has $\{1.006 \quad 0.868 \quad 0.726 \quad 1.093\}$ as its diagonal elements.

■

The relationship between RGA, NI, and integrity is further explored in the following corollary in (Chiu and Arkun 1990).

Corollary 2.1

For $m > 2$, and any λ_{ii} negative, the decentralized feedback control structure has one of the following properties:

1. If $\mathrm{NI}[G(0)] < 0$, the closed loop plant is unstable, but the reduced plant without the ith loop can be stabilized.
2. If $\mathrm{NI}[G(0)] > 0$, the closed loop plant is unstable after the failure of the corresponding i^{th} loop.

Proof

1. Using the result of theorem 2.2 if $\mathrm{NI}(G(0)) < 0$, the closed loop plant is unstable. But, using the equation (2.43) the $\mathrm{NI}(G^{ii}(0))$ will be positive and so the reduced plant without the ith loop can be stabilized.
2. Also, if $\mathrm{NI}(G(0)) > 0$ then the $\mathrm{NI}(G^{ii}(0))$ will be negative and so the reduced plant without the ith loop will be unstable. Hence, the closed loop plant after the failure of the corresponding i^{th} loop is unstable.

■

Example 2.2.10

Consider the following steady state transfer function matrix (Niederlinski 1971)

$$G(0) = \begin{bmatrix} 0.2 & 0.8 & 0.3 \\ -1.0 & 0.1 & 1.0 \\ 0.5 & -0.6 & 0.1 \end{bmatrix}$$

and the corresponding RGA as

$$\Lambda = \begin{bmatrix} 0.159 & 0.626 & 0.215 \\ 0.339 & -0.017 & 0.678 \\ 0.502 & 0.391 & 0.107 \end{bmatrix}$$

The diagonal elements of the RGA are $\{0.159 \quad -0.017 \quad 0.107\}$ and the NI is 383.50. Then, using corollary 2.1 the closed loop plant will be unstable, if the second loop fails. Therefore, the pairing $(u_1 - y_1, u_2 - y_2, u_3 - y_3)$ is not recommended by the RGA, NI and integrity considerations. However, by permuting the 1st and 2nd row, we have

$$G(0) = \begin{bmatrix} -1.0 & 0.1 & 1.0 \\ 0.2 & 0.8 & 0.3 \\ 0.5 & -0.6 & 0.1 \end{bmatrix}$$

The diagonal elements of RGA are now $\{0.339 \quad 0.626 \quad 0.107\}$ and the corresponding NI is 9.587. Therefore, using theorem 2.2 the close loop plant and the reduced plants can be stable and the pairing $(u_1 - y_2, u_2 - y_1, u_3 - y_3)$ is recommended.

■

In industrial control systems, integrity issues arise in sensor or actuator failure. These failures can be categorized as: Hardware failure or variable saturation. Variable saturation is a very common phenomenon in industrial applications. Whenever a valve is fully opened, the electrical machine is working under full load condition, heating or cooling devices are working under maximum load, variable saturation is inevitable. Hardware failure occurs for example in:

- Failure in machine control centre for electrical motors in different actuator applications.
- Loss of oil pressure in hydraulic actuators.
- Loss of air pressure and air leakage in pneumatic actuators.
- Mechanical failure in rotary flow sensors.
- Burned out thermocouple.
- Material precipitation in pH sensors.

Sensor and actuator failures are both detrimental to control systems. Sending a wrong signal to the controller, in the case of sensor failure, or terminating the control action in one loop, in the case of actuator failure, can both lead to serious control problems and instability. However, in an industrial environment such faults are promptly detected and the controller in the failure loop is brought out of service and the failed part is replaced or repaired. Meanwhile, it is necessary to guarantee the closed loop stability to minimize the damage.

The perseverance of closed loop stability in the face of sensor or actuator failure may be achieved if the multivariable plant with all its loops closed and the plant with the jth sensor or actuator removed, i.e. $k_j = 0$ is integral stabilizable. A direct consequence of theorem 2.1 gives the condition under which this desirable property does not exist.

Theorem 2.3
Let $P(s)$ be rational, proper and integral stabilizable. Then, the closed loop plant with the jth sensor or actuator failed is not integral stabilizable if $\det(P^{jj}(0)) < 0$.

■

This theorem clearly shows the cases that with the failure of the jth sensor or actuator, the closed loop plant with integral control will become unstable.

Following the results of (Grosdidier et al. 1985), we can now establish a relationship between integrity and the RGA.

Theorem 2.4
If,

- $\lambda_{ij}(0) < 0$.
- $C(s)$ is a diagonal compensator matrix.
- $P(s)$ is proper.

Then, with any positive gain $k > 0$, the closed loop plant shown in figure 2.1 has at least one of the following properties:

- It is unstable.
- The jth loop, corresponding to $(u_j - y_i)$, is unstable with all other loops opened.
- The closed loop plant is unstable with the jth loop removed.

Proof. The third RGA property stated that its elements are invariant under input and output scaling. As the diagonal compensation can be regarded as a scaling matrix, we have

$$\lambda_{ij} = (-1)^{i+j} g_{ij}(0) \frac{\det(G^{ij}(0))}{\det(G(0))} \tag{2.44}$$

and

$$\lambda_{ij} = (-1)^{i+j} p_{ij}(0) \frac{\det(P^{ij}(0))}{\det(P(0))} \tag{2.45}$$

If $\lambda_{ij} < 0$, then one or all of the terms in equation (2.45) will be negative. For $\det(P(0)) < 0$, the first property will hold. For $p_{ij}(0) < 0$, the second property will hold. For $\det(P^{ij}(0)) < 0$, the third property will hold. This concludes the proof.

■

Note 1. If any of the diagonal elements of the RGA is negative, decentralized control with integral action will lead to a closed loop plant with no integrity. This is not at all recommended.

Note 2. In 2×2 multivariable plants, it is always possible to change the plant pairing to obtain positive diagonal elements for the corresponding RGA. However, this is not the case for plants with number of inputs and outputs greater than or equal to three.

Note 3. In theorem 2.4, the condition of negative diagonal elements of the RGA is sufficient but not necessary. That is for plants with number of inputs and outputs greater than or equal to three, and with positive RGA diagonal elements, all properties of theorem 2.4 may hold.

Example 2.2.11

Consider the steady state transfer function matrix as

$$G = \begin{bmatrix} 1.0 & 1.0 & -0.1 & 0.1 \\ \alpha & 2.0 & -1.0 & 0.2 \\ -2.0 & -3.0 & 1.0 & 1.0 \\ -0.1 & 0 & 0.1 & 1.0 \end{bmatrix}$$

For $\alpha = 1$ the corresponding RGA is

$$\Lambda = \begin{bmatrix} 3.77 & -3.05 & 0.22 & 0.06 \\ 1.71 & -4.19 & 3.38 & 0.10 \\ -4.79 & 8.24 & -2.99 & 0.54 \\ 0.31 & 0 & 0.39 & 0.3 \end{bmatrix}$$

and interchanging the second and third rows gives $\lambda_{ii}(0) > 0$. For $\alpha = 0.1$ the RGA is

$$\Lambda = \begin{bmatrix} -7.04 & 10.22 & -1.92 & -0.26 \\ -0.32 & 7.82 & -6.31 & -0.19 \\ 8.94 & -17.04 & 10.61 & -1.51 \\ -0.58 & 0 & -1.38 & 2.96 \end{bmatrix}$$

Obviously, in this case no permutation can make $\lambda_{ii}(0) > 0$ for all i.

■

Following the integrity property, another requisite called the *integral controllability* (IC) is proposed. See (Grosdidier et al. 1985). The integral stabilizability definition only ensures the existence of a constant $k > 0$ such that the closed loop plant shown in figure 2.1 is stable. It is clearly desirable for the closed loop plant to maintain stability in the case of gain reduction towards zero.

Definition 2.2. The open loop stable plant $P(s)$ is *integral controllable* if there exists a $k^* > 0$ such that the closed loop plant shown in figure 2.1 is stable for all $k \in (0, k^*]$, and has zero steady state error.

In practice, integral controllable plants can be tuned by increasing the gain from a small value that ensures stability until acceptable performance is achieved. The following theorem proved in (Grosdidier et al. 1985) provides the conditions for integral controllability.

Theorem 2.5
The rational plant $P(s)$ is integral controllable if all the eigenvalues of $P(0)$ lie in the open right half complex plane. And, it is not integral controllable if any of the eigenvalues of $P(0)$ lie in the open left half complex plane.

■

Note 1. Theorem 2.5 does not consider the cases with $P(0)$ eigenvalues on the imaginary axis, including zero.

Note 2. For integral stabilizability we require that $\det(P(0)) > 0$. This can occur with an even number of left half complex plane eigenvalues, which clearly violates the integral controllability condition. However, in the case of SISO plants these would be equivalent.

Example 2.2.12

Consider the plant

$$P(s) = \begin{bmatrix} \dfrac{-3(-s+1)}{(s+1)(0.5s+1)} & \dfrac{4}{(0.5s+1)} \\ \dfrac{-4}{(0.5s+1)} & \dfrac{2}{(0.5s+1)} \end{bmatrix}$$

where the eigenvalues of $P(0)$ are $\{-0.5 \pm j3.12\}$, and the plant is not integral controllable. However, $\det(P(0)) = 10 > 0$ and it may be integral stabilizable. To check the integral stabilizability of the plant, its characteristic equation is derived. The closed loop characteristic equation is

$$s^5 + 5s^4 + (10k + 8)s^3 + (18k + 4)s^2 + (88k^2 - 4k)s + 40k^2 = 0$$

and it can easily be shown that the closed loop plant is stable for $0.157 < k < 0.389$. The multivariable plants which are stable for a gain interval (excluding zero gain) are called *conditionally integral controllable* plants. This is an undesirable situation in practical control systems, since the exact lower and upper bounds may be difficult to find and may be subject to variations. Also, integrity is lost in such plants.

■

The following theorem can be easily derived from theorem 2.5 regarding the important issue of sensor and actuator failures mentioned earlier in this section.

Theorem 2.6
Let $P(s)$ be rational, proper and integral controllable. Then, the closed loop plant with the jth sensor (actuator) removed is integral controllable if all the eigenvalues of $P^{jj}(0)$ are in the open right half complex plane. It is not integral controllable with the jth sensor (actuator) removed if any of the eigenvalues of $P(0)$ or $P^{jj}(0)$ lie in the left half complex plane.

■

Example 2.2.13

Consider the closed loop plant shown in figure 2.1 with $C(s) = I$, and

$$P(0) = \begin{bmatrix} 3 & 1 & 1 \\ -2 & 1 & 3 \\ 1 & 1 & 2 \end{bmatrix}$$

where the eigenvalues of $P(0)$ are $\{0.1206, 2.347, 3.532\}$, and the plant is integral controllable. If loop 1 fails, the reduced plant will be

$$P^{11}(0) = \begin{bmatrix} 1 & 3 \\ 1 & 2 \end{bmatrix}$$

the eigenvalues of $P^{11}(0)$ are $\{-0.303, 3.303\}$, and is not integral controllable, due to its negative eigenvalue. If loop 2 fails, the reduced plant will be

$$P^{22}(0)=\begin{bmatrix}3 & 1\\ 1 & 2\end{bmatrix}$$

the eigenvalues of $P^{22}(0)$ are $\{1.382, 3.618\}$, and is integral controllable. If loop 3 fails, the reduced plant will be

$$P^{33}(0)=\begin{bmatrix}3 & 1\\ -2 & 1\end{bmatrix}$$

the eigenvalues of $P^{33}(0)$ are $\{2 \pm j1\}$, and is integral controllable. Therefore, in the face of actuator or sensor failures in the second or third loop the closed loop plant will remain stable, but failure in the first loop leads to closed loop instability.

■

Apart from 2×2 plants and a weak result for 3×3 plants, unlike the established relationship between integrity and the RGA developed in theorem 2.4, the RGA does not provide information on the sensor or actuator failures as given in theorem 2.6. However, the NI can be used to study the failure situations. It is shown in (Chiu and Arkun 1990) that if $\mathrm{NI}[G^{ii}(0)] < 0$, the reduced plant after an actuator or sensor failure is not stabilizable. Also, the following corollary in (Chiu and Arkun 1990) provides a necessary condition for closed loop integrity in the face of actuator or sensor failures using the NI.

Corollary 2.2
The closed multivariable plant under decentralized control can be stabilized and remain stable after the failure of any j^{th} loop only if

1. $\mathrm{NI}[G(0)] > 0$, and
2. $\lambda_{jj}((G(0)) > 0$ for all $j = 1,\ldots,m$.

■

Theorem 2.7
Let $G(s)$ be a 2×2 plant. Then, if both the diagonal elements of its RGA are positive, there exists a diagonal compensator such that the closed loop plant with the jth sensor (actuator) ($j = 1,2$) removed, is integral controllable.

■

Note 1. In the case of 3×3 plants, the closed loop plant with the jth sensor (actuator) ($j = 1,2,3$) removed is integral controllable, if there exists a diagonal compensator such that $P(s)$ is integral controllable and $p_{jj}(0) > 0$ for $j = 1,2,3$.

Note 2. If the compensator matrix is chosen as $C(0) = G^{-1}(0)$, then $P(0) = I$ and it possess all the requirements of theorem 2.6. Also, the controller can be tuned on line to achieve the desired performance. In (Porter and Khaki-Sedigh 1989), the robustness property of such PI controllers is derived.

Finally, in this section we consider the notion of *Decentralized Integral Controllability* (DIC). A multivariable plant is *Decentralized Integral Controllable* if it is possible to design a diagonal controller as shown in figure 2.1, such that:

- All individual loops are stable.
- With the simultaneous closure of the loops, the multivariable plant remains stable.
- Each loop gain can be independently detuned to zero without jeopardizing the closed loop stability.

Following the derivation in (Campo and Morari 1994) the results for DIC are now presented. First, we present some definitions. Let the index set T consists of k tuples of integers in the range $1,\cdots,m$ and be defined by

$$\mathrm{T} \triangleq \left\{(i_1,\ldots,i_k) \mid \ 1 \le i_1 < i_2 < \cdots < i_k \le m\right\} \tag{2.46}$$

Then, for each $t = \left(i_1,\cdots,i_k\right) \in \mathrm{T}$ a $k \times k$ principal sub-matrix of G denoted by G_t is defined, that is constructed by the rows and columns of $G(s)$ induced by t.

The real matrix G is called *D-Stable* if and only if for all diagonal matrices $D > 0$, the eigenvalues of GD lie in the RHP. Also, define the sign matrix as

$$S_G \triangleq diag\left\{sign(g_{11}),\ldots,sign(g_{mm})\right\}$$

where

$$sign(x) = \begin{cases} 1 & \text{if} \quad x > 0 \\ -1 & \text{if} \quad x < 0 \end{cases}$$

Then, $G^+ = GS_G$.

Definition 2.3. The multivariable plant $G(s)$ is DIC if there exists a controller $\frac{1}{s}C(s)$, as shown in figure 2.1, such that the closed loop plant is stable for all $E\frac{1}{s}C(s)$, where $E \in \varepsilon_D$ and $\varepsilon_D = \left\{E = diag(\varepsilon_i) \middle| \varepsilon_i \in [0,1], i = 1,\cdots,m\right\}$.

Note. The stability condition in definition 2.3 is also called *decentralized unconditional stability*. This means that the gain of each controller can be detuned to zero or vary in the range [0,1], independently. Also, the closed loop plant is called *unconditionally stable* if $E\frac{1}{s}C(s)$ stabilizes $G(s)$ for all $E \in \varepsilon_I$, where $\varepsilon_I = \{E = kI \mid k \in (0,1]\}$. This is similar to the integral controllability definition 2.2.

Theorem 2.8

The multivariable plant $G(s)$ is DIC only if the eigenvalues of $G_t^+(0)D$ for all t and all diagonal $D > 0$ lie in the closed RHP less the origin.

Proof. It is easily observed from the definition of decentralized unconditional stabilizability that the closed loop plant shown in figure 2.1 is DIC if and only if there exists $\frac{1}{s}C(s)$ with $C(s)$ diagonal and $C(0)$ nonsingular such that $\frac{1}{\|D\|_2}[D\frac{1}{s}C(s)]_t$ unconditionally stabilizes $G_t(s)$ for all $t \in \mathrm{T}$ and all $D > 0$. On the other hand, this implies using theorem 2.5 that all the eigenvalues of $\frac{1}{\|D\|_2}[G(0)DC(0)]_t$ must lie in the closed RHP less the origin for all $t \in \mathrm{T}$ and all $D > 0$. Therefore, all the eigenvalues of $[G(0)DC(0)]_t$ must lie in the closed RHP less the origin for all $t \in \mathrm{T}$ and all $D > 0$. By writing

$$G(0) = G^+(0)S_{G(0)}^{-1} = G^+(0)S_{G(0)}$$

$$C(0) = C^+(0)S_{C(0)}^{-1} = S_{C(0)}C^+(0)$$

We have that all the eigenvalues of $[G^+(0)S_{G(0)}S_{C(0)}C^+(0)D]_t$ must lie in the closed RHP less the origin for all $t \in \mathrm{T}$ and all $D > 0$. Note that for $D = I$ and $\dim(t) = 1$ we have $g_{ii}(0)c_{ii}(0) > 0$ for all $i \in \{1,\cdots,m\}$. Hence, $S_{G(0)}S_{C(0)} = I$. Therefore, all the eigenvalues of $[G^+(0)C^+(0)D]_t$ must lie in the closed RHP less the origin for all $t \in \mathrm{T}$ and all $D > 0$. However, $C^+(0)D$ is an arbitrary positive definite matrix. Hence, all the eigenvalues of $G_t^+(0)D$ must lie in the closed RHP less the origin for all $t \in \mathrm{T}$ and all $D > 0$.

■

Other necessary condition results for DIC of $G(s)$ can be found in (Campo and Morari 1994). The next corollary is from (Morari and Zafiriou 1989).

Corollary 2.3

The multivariable plant $G(s)$ is DIC only if both of these conditions hold,

1. All the eigenvalues of $G^+(0)D$ lie in the closed RHP for all $D \geq 0$.
2. The relative gains of $G(s)$ are positive.

Proof. Condition 1 is directly proved from theorem 2.8. Also from theorem 2.6 we have that DIC of $G(s)$ implies that all the eigenvalues of $G_t^+(0)$ must lie in the closed RHP less the origin for all $t \in \mathrm{T}$. It can be shown that this implies all the relative gains of $G^+(s)$ must be positive (Campo and Morari 1994). And, it then follows that the relative gains of $G(s)$ must be positive.

■

Note 1. DIC is a stronger and more desirable property than IC. For IC, all gains must be simultaneously reduced to zero, while for DIC loop gains can decrease at different rates. Hence, a multivariable plant can be IC and not DIC. But a DIC multivariable plant is necessarily IC. This implies that the IC test can be performed prior to the DIC test, and remove the possible pairing selections that fail the IC test.

Note 2. DIC is an inherent multivariable plant property and IC is dependent on the designed decentralized controller. This is clearly observed form the relevant theorem presented for the DIC and IC. For more details see (Skogestad and Morari 1992).

2.3 Pairing Rules

Based on the above properties, we will now present the input-output pairing rules with the RGA. The following points are crucial to choose the input-output pair:

- Choose the input-output pair with corresponding RGA element close to one (Bristol 1966).
- Niederlinski Index must be positive (Niederlinski 1971), (Chiu and Arkun 1991).
- RGA elements corresponding to the input-output pair must be positive.
- Larger RGA elements are not appropriate for input-output pairing (Skogestad and Morari 1987).

The following worked examples show the application of the above rules to some benchmark plants.

Example 2.3.1

Consider the Quadruple-tank plant shown in figure 2.2 (Johansson 2000). The control objective is the control of two lower tanks levels, h_1 and h_2, using the two pumps. Input voltages to the pumps are v_1 and v_2. Using the mass balances and Bernoulli's law, the equations describing the plant are

$$\dot{h}_1 = -\frac{a_1}{A_1}\sqrt{2gh_1} + \frac{a_3}{A_1}\sqrt{2gh_3} + \frac{\gamma_1 k_1}{A_1}v_1$$

$$\dot{h}_2 = -\frac{a_2}{A_2}\sqrt{2gh_2} + \frac{a_4}{A_2}\sqrt{2gh_4} + \frac{\gamma_2 k_2}{A_2}v_2$$

$$\dot{h}_3 = -\frac{a_3}{A_3}\sqrt{2gh_3} + \frac{(1-\gamma_2)k_2}{A_3}v_2$$

$$\dot{h}_4 = -\frac{a_4}{A_4}\sqrt{2gh_4} + \frac{(1-\gamma_1)k_1}{A_4}v_1$$

where h_i, A_i and a_i for $i = 1,\dots,4$ denote the level, cross-section and outlet hole cross sections of the ith tank, respectively. Also, g denotes the gravity acceleration. In the above equations, $k_i v_i$ is the corresponding flow to the pump voltage, v_i. Moreover, $\gamma_1, \gamma_2 \in (0,1)$ indicate the division ratio of flow using the valves.

If we define the manipulated variables as $u_1 = v_1$, $u_2 = v_2$ and the measured variables as $y_1 = h_1$, $y_2 = h_2$, the transfer function model of the Quadruple-tank is

$$G(s) = \begin{bmatrix} \dfrac{\gamma_1 T_1 k_1 k_c}{A_1(sT_1+1)} & \dfrac{(1-\gamma_2)T_1 k_2 k_c}{A_1(sT_1+1)(sT_3+1)} \\ \dfrac{(1-\gamma_1)T_2 k_1 k_c}{A_2(sT_2+1)(sT_4+1)} & \dfrac{\gamma_2 T_2 k_2 k_c}{A_2(sT_2+1)} \end{bmatrix}$$

where

$$T_i = \frac{A_i}{a_i}\sqrt{\frac{2h_i^0}{g}}$$

and h_i^0 are the operating water levels. The corresponding RGA is

$$\Lambda = \begin{bmatrix} \lambda & 1-\lambda \\ 1-\lambda & \lambda \end{bmatrix}$$

where $\lambda = \dfrac{\gamma_1\gamma_2}{\gamma_1+\gamma_2-1}$. If $\lambda > 0$ or $1 < \gamma_1 + \gamma_2 < 2$ then $(u_1 - y_1, u_2 - y_2)$ is the appropriate pair and the corresponding Niederlinski index is $\dfrac{1}{\lambda}$, which is positive.

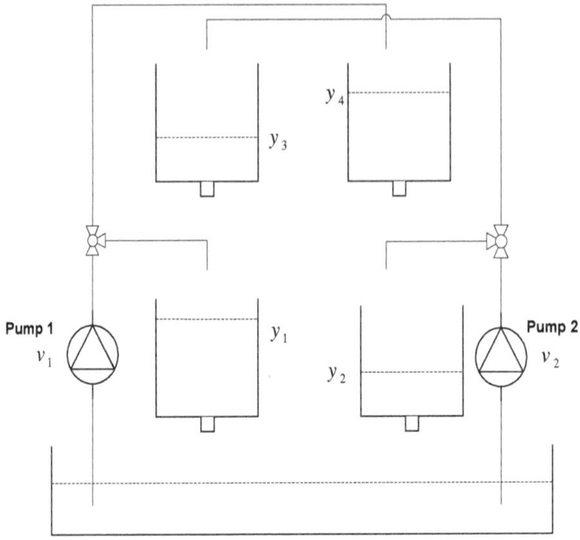

Fig. 2.2 The Quadruple-tank.

However, for $\gamma_1+\gamma_2<1$, (u_1-y_2, u_2-y_1) is the appropriate input-output pair with $\mathrm{NI}[G(0)]=\dfrac{1-\gamma_1-\gamma_2}{(1-\gamma_1)(1-\gamma_2)}=\dfrac{1}{1-\lambda}>0$. Therefore, according to corollary 2.2, decentralized control can be used to stabilize the plant with integrity.

■

Example 2.3.2

Consider the linear multivariable plant introduced by Hovd and Skogestad (1992) as

$$G(s)=\frac{1-s}{(5s+1)^2}\begin{bmatrix}1 & -4.19 & -25.96\\ 6.19 & 1 & -25.96\\ 1 & 1 & 1\end{bmatrix}$$

where its corresponding RGA is

$$\Lambda=\begin{bmatrix}1.00 & 5.00 & -5.00\\ -5.00 & 1.00 & 5.00\\ 5.00 & -5.00 & 1.00\end{bmatrix}$$

Hence, $(u_1-y_1, u_2-y_2, u_3-y_3)$ is the appropriate input-output pair proposed by the RGA. Hovd and Skogestad (1992) proposed two sets of optimized PI controllers, as shown in table 2.1, for the pairing $(u_1-y_1, u_2-y_2, u_3-y_3)$ related to

$\lambda_{ij}=1$ and for the pairing $(u_1-y_3,u_2-y_1,u_3-y_2)$ related to $\lambda_{ij}=5$. Figure 2.3 shows the closed loop step responses in both cases. It is obvious that the closed loop step response performance corresponding to $\lambda_{ij}=5$ is significantly better than that of $\lambda_{ij}=1$. This clearly shows that the RGA pairing rules are not necessarily optimal.

■

Example 2.3.3

Consider the following multivariable transfer function matrix (Mc Avoy et al. 2003)

$$G(s)=\begin{bmatrix}\frac{5e^{-40s}}{100s+1} & \frac{e^{-4s}}{10s+1}\\ \frac{-5e^{-4s}}{10s+1} & \frac{5e^{-40s}}{100s+1}\end{bmatrix}$$

Table 2.1 Parameters of the PI Controllers as $k_i(1+\frac{1}{\tau_i s})$.

	$\lambda_{ij}=1$ $(u_1-y_1,u_2-y_2,u_3-y_3)$		$\lambda_{ij}=5$ $(u_1-y_3,u_2-y_1,u_3-y_2)$	
	k_i	τ_i	k_i	τ_i
$i=1$	0.1230	32.40	-0.6840	24.15
$i=2$	0.1443	34.54	-0.02425	7.270
$i-3$	0.002940	3.988	0.007685	0.3688

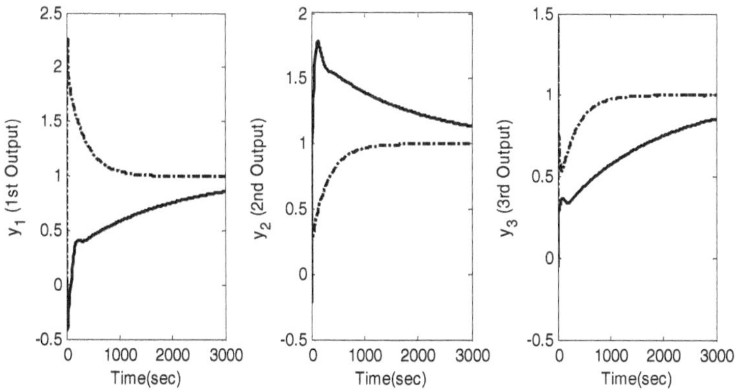

Fig. 2.3 Closed loop step responses under decentralized control in example 2.3.2, solid line corresponds to $\lambda_{ij}=1$ and Dash-dot line corresponds to $\lambda_{ij}=5$.

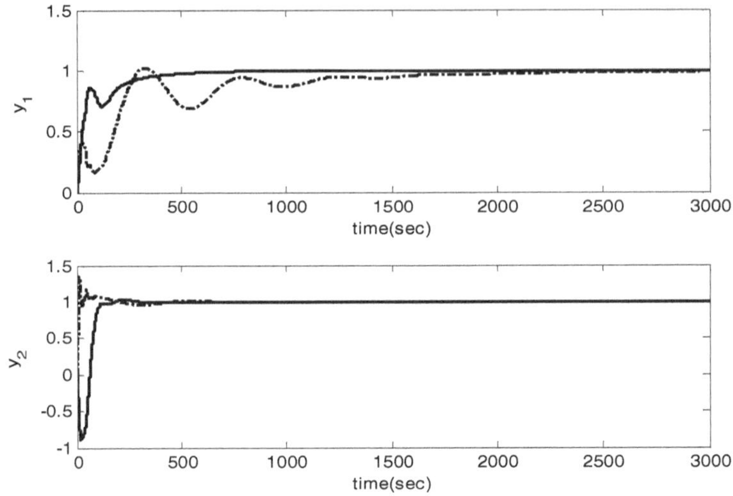

Fig. 2.4 Closed loop step responses under decentralized control in example 2.3.3, solid line corresponds to diagonal and Dash-dot line corresponds to off diagonal pairing.

The corresponding RGA is

$$\Lambda = \begin{bmatrix} 0.8333 & 0.1667 \\ 0.1667 & 0.8333 \end{bmatrix}$$

The proposed input-output pair is $(u_1 - y_1, u_2 - y_2)$. However, the other possible pair is $(u_1 - y_2, u_2 - y_1)$. To analyze the performance of the decentralized control corresponding to these pairings, two optimized PI controllers are designed in (Mc Avoy et al. 2003). The closed loop step responses shown in figure 2.4 clearly show the advantage of diagonal pairing proposed by the RGA method.

■

2.4 Control Configuration Selection of Nonsquare and Unstable Multivariable Plants

In this section, the RGA analysis developed in the preceding sections is extended to multivariable plants with an unequal number of inputs and outputs, i.e. nonsquare plants, and open loop unstable multivariable plants.

2.4.1 Control Configuration Selection of Nonsquare Multivariable Plants

Nonsquare plants are defined as multivariable plants with an unequal number of inputs and outputs. Nonsquare plants incorporate the two cases: more inputs than outputs and more outputs than inputs. Although nonsquare plants are often encountered in many engineering disciplines, e.g. chemical processes, the theory for analysis and control of nonsquare plants is not as developed as one would expect. The main control strategy for nonsquare plants is the *squaring down* process. That is, necessary number of outputs or inputs are added or deleted from the transfer function matrix to obtain a *square* plant. Then, the well established control paradigms can be applied to the nonsquare plants. However, both of these solutions have their own related problems. Adding unnecessary outputs and inputs means more costs and maintenance issues and deleting the manipulated inputs will reduce the degrees of freedom for achieving the desired responses and deleting outputs result in less reliable measured information regarding the plant performance.

In the case of nonsquare plants with more inputs than outputs, there are reports indicating the superior performance of using the nonsquare structure in comparison with the squared plant. See for example, (Treiber and Hoffman 1986), and (Morari et al. 1985). In this case, there is no specific control difficulty and making some actuators redundant has less deteriorating affects on the closed loop plant performance. However, nonsquare plants with more outputs than inputs are *functionally uncontrollable* and the outputs cannot be *perfectly* derived towards the desired set points. Square plants with one or more actuators saturated and continually kept saturated, are examples of such nonsquare plants. In this case, the notion of *perfect control in the least squares sense* is proposed. That is, the sum of the squared output errors is minimized and the control will be perfect in a least square sense. This can be considered as a generalization of perfect control to nonsquare plants with more outputs than inputs.

2.4.1.1 The Nonsquare Relative Gain

Control configuration selection of nonsquare multivariable plants is dealt with in (Reeves and Arkun 1989) and (Chang and YU 1990). The material in this section is based on (Chang and YU 1990).

Consider the linear multivariable plant described by the following transfer function matrix model

$$Y(s) = G(s)U(s) \tag{2.47}$$

where $G(s)$ is a $l \times m$ matrix of rational functions, $Y(s)$ is the $l \times 1$ output vector, and $U(s)$ is the $m \times 1$ input vector. In the case of nonsquare plants $l > m$, the idea of *tight* or *perfect* control proposed in section 2.2.1 to derive the RGA definition is not possible. This is easily observed from the lack of functional controllability of the plant. Hence, the notion of perfect control is modified to cope with this case. This modification entails the minimization of the steady state errors in a least squares sense. That is, a controller is designed to minimize the sum of the squares of the error vector. This controller is called a *least square tight or perfect control*. Let us consider the standard Internal Model Control (IMC) structure shown in figure 2.5.

Let, $\tilde{G}(s)$ be the plant model, $G_c(s)$ the controller transfer function matrix, and $R(s)$ the reference input. Then, the following equations govern the closed loop plant

$$Y_a(s) = Y(s) - \tilde{G}(s)U(s) \tag{2.48}$$

$$U(s) = G_c(s)(R(s) - Y_a(s)) \tag{2.49}$$

where $Y_a(s)$ is the augmented output shown in figure 2.5. The closed loop plant transfer function matrix is derived from equations (2.47), (2.48) and (2.49)

$$Y(s) = G(s)[I_m + G_c(s)(G(s) - \tilde{G}(s))]^{-1}G_c(s)R(s) \tag{2.50}$$

The closed loop tracking error is obtained as follows. Let, the tracking error be defined as, $E(s) = R(s) - Y(s)$. Then

$$E(s) = \left[I_l - G(s)[I_m + G_c(s)(G(s) - \tilde{G}(s))]^{-1}G_c(s)\right]R(s) \tag{2.51}$$

which in the case of perfect plant modeling, i.e. $G(s) = \tilde{G}(s)$, gives

$$E(s) = [I_l - G(s)G_c(s)]R(s) \tag{2.52}$$

To get the minimum steady state error in the least square sense, tight control would require that the steady state controller transfer function matrix be determined from the pseudo inverse of the steady state plant gain matrix as

$$G_c(0) = (G^T(0)G(0))^{-1}G^T(0) = G^{\dagger}(0) \tag{2.53}$$

where $\dagger$ denotes the Moore-Penrose pseudo inverse. Hence, the final value theorem gives the steady state error and the control vector as

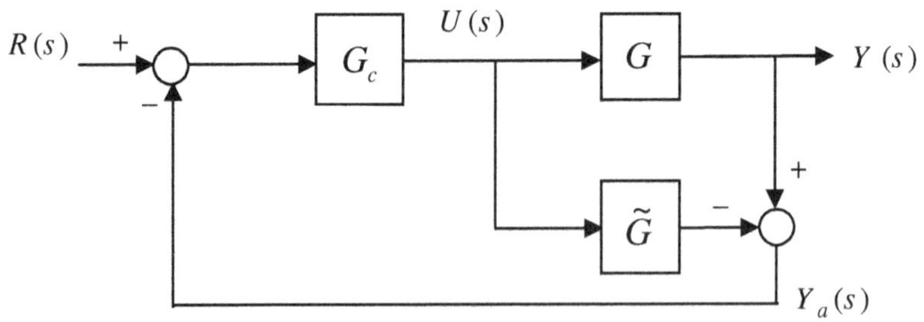

Fig. 2.5 The IMC structure.

$$E(0) = [I_l - G(0)G^{\dagger}(0)]R(0)$$
$$U(0) = G^{\dagger}(0)R(0) \tag{2.54}$$

Similar to the arguments in section 2.2.1, if the condition of tight control in the least square sense is satisfied, we have

$$\delta u_j = g^{\dagger}_{ji}\,\delta y_i \tag{2.55}$$

where $g^{\dagger}_{ji}$ is the $(j,i)^{th}$ element of $G^{\dagger}(0)$. And, it is now possible to define the *nonsquare relative gain* (NSRG) as

$$\lambda^N_{ij} = g_{ij} \cdot g^{\dagger}_{ji} \tag{2.56}$$

and the *nonsquare relative gain array* (NSRGA) is

$$\Lambda^N = [\lambda^N_{ij}] = G \cdot * (G^{\dagger})^T \tag{2.57}$$

This is similar to the definition given by equation (2.18).

Example 2.4.1

Consider a side stream distillation column with the following transfer function matrix (Chang and Yu 1990)

$$G(s) = \begin{bmatrix} \dfrac{-9.811e^{-1.59s}}{11.36s+1} & \dfrac{0.374e^{-7.75s}}{(22.2s+1)^2} & \dfrac{-11.3e^{-3.79s}}{(21.74s+1)^2} \\ \dfrac{5.984e^{-2.24s}}{14.29s+1} & \dfrac{-1.986e^{-0.71s}}{(66.67s+1)^2} & \dfrac{5.24e^{-60s}}{(400s+1)^2} \\ \dfrac{2.38e^{-0.42s}}{(1.43s+1)^2} & \dfrac{0.0204e^{-0.59s}}{(7.14s+1)^2} & \dfrac{-0.33e^{-0.68s}}{(2.38s+1)^2} \\ \dfrac{-11.67e^{-1.91s}}{12.19s+1} & \dfrac{-0.176e^{-0.48s}}{(6.9s+1)^2} & \dfrac{4.48e^{-0.52s}}{(11.11s+1)^2} \end{bmatrix}$$

the corresponding steady state gain of the plant is

$$G(0) = \begin{bmatrix} -9.811 & 0.374 & -11.3 \\ 5.984 & -1.986 & 5.24 \\ 2.38 & 0.0204 & -0.33 \\ -11.67 & -0.176 & 4.48 \end{bmatrix}$$

and its pseudo inverse is

$$G^{\dagger}(0)=\begin{bmatrix} -0.0245 & 0.0010 & 0.0117 & -0.0622 \\ -0.2748 & -0.5509 & 0.0205 & -0.0473 \\ -0.0762 & -0.0191 & -0.0060 & 0.0530 \end{bmatrix}$$

Hence, the NSRGA can be computed using (2.57) as

$$\Lambda^{N} = G(0).*(G^{\dagger}(0))^{T} = \begin{bmatrix} 0.241 & -0.103 & 0.861 \\ 0.006 & 1.094 & -0.100 \\ 0.028 & 0.000 & 0.002 \\ 0.726 & 0.008 & 0.237 \end{bmatrix}$$

■

2.4.1.2 The Nonsquare Relative Gain Array Properties

The NSRGA properties are derived by Chang and YU (1990). A summary of these properties are given below:

Property 1. *Summation of the elements of each column of the NSRGA is unity.* This property is consistent with the corresponding RGA property in section 2.2.2.

Property 2. *Summation of the elements of each row of the NSRGA is between zero and unity.* This is a direct consequence of the loss of zero steady state tracking error in the nonsquare plant with more outputs than inputs. It is also the fundamental difference between the NSRGA and the RGA.

Property 3. *The NSRGA is input scaling independent but output scaling dependent.* This implies that in contrary to the RGA, the NSRGA is variant under output scaling. This is interpreted in a weighted least square sense, which weights the relative importance of the different outputs in the least square minimization.

Property 4. *Any permutation of rows and columns in plant transfer function matrix results in the same permutation in the NSRGA.*

Property 5. *For square multivariable plants* $\Lambda^{N}(G)=\Lambda(G)$.

Property 6. *For single-input multi-output multivariable plants (SIMO), we have* $G(s)=\begin{bmatrix} g_{11}(s) & g_{21}(s) & \cdots & g_{l1}(s)\end{bmatrix}^{T}$ *and the NSRGA is* $\Lambda^{N}=\begin{bmatrix}\lambda_{11}^{N} & \lambda_{21}^{N} & \cdots & \lambda_{l1}^{N}\end{bmatrix}^{T}$, *where* $\lambda_{i1}^{N}=\dfrac{g_{i1}^{2}}{\sum_{k=1}^{l} g_{k1}^{2}}$.

Property 7. *The NSRGA elements tend to infinity as the nonsquare plant transfer function matrix approaches singularity.*

Example 2.4.2

Consider the following steady state gain of a nonsquare plant

$$G = \begin{bmatrix} a & a \\ a & a \\ a & a(1+\delta) \end{bmatrix}$$

This plant is nearly singular. It is singular for $\delta = 0$, where $\det(G^T G) = 0$. This is the necessary and sufficient condition for the singularity of a nonsquare matrix. The corresponding NSRGA is

$$\Lambda^N = \begin{bmatrix} \frac{1+\delta}{2\delta} & \frac{-1}{2\delta} \\ \frac{1+\delta}{2\delta} & \frac{-1}{2\delta} \\ \frac{-1}{\delta} & \frac{1+\delta}{\delta} \end{bmatrix}$$

For, $\delta = 0.001$ and $\delta = 0.0001$, the corresponding NSRGAs are

$$\Lambda_1^N = \begin{bmatrix} 500.5 & -500 \\ 500.5 & -500 \\ -1000 & 1001 \end{bmatrix} \text{ and, } \Lambda_2^N = \begin{bmatrix} 5000.5 & -5000 \\ 5000.5 & -5000 \\ -10000 & 10001 \end{bmatrix}$$

■

It should be noted that singularity of the nonsquare plant rarely occurs in practice. Singularity of square plants is more common. This is partly because the singularity of a nonsquare matrix requires the simultaneous vanishing of many minors depending on the dimension of the matrix. But, in a square plant the determinant of the matrix must be zero. Exactly for the same reason, it is also well known that transmission zeros for nonsquare plants are very uncommon.

Property 8. *Relative changes in the elements of the nonsquare plant* g_{ij} *and its NSRGA elements* λ_{ij}^N *are related by the following equations*

$$\frac{d\lambda_{ij}^N}{\lambda_{ij}^N} = \left[\frac{g_{ij}^2 \det[(G^{ij})^T (G^{ij})]}{\lambda_{ij}^N \det(G^T G)} + (1 - 2\lambda_{ij}^N) \right] \frac{dg_{ij}}{g_{ij}} \tag{2.58}$$

This result can show that a plant with larger λ_{ij}^{N} is more sensitive to errors in g_{ij}.

Example 2.4.3

Consider the following steady state gain of a nonsquare plant

$$G(0)=\begin{bmatrix} 9.77 & 7.53 \\ 13 & 10 \\ 13.6 & 10.85 \end{bmatrix}$$

where its NSRGA is

$$\Lambda^{N}=\begin{bmatrix} 9.42 & -9.09 \\ 19 & -18.33 \\ -27.42 & 28.42 \end{bmatrix}$$

Note that the (1,1) element of the NSRGA is smaller than the (2,1) element. Hence, using property 8, it should be less sensitive to perturbation in the corresponding element of the transfer function matrix. We perturb the elements of $G(0)$ by 10% in the following two cases:

$$G_{p1}(0)=\begin{bmatrix} 9.77(1+0.1) & 7.53 \\ 13 & 10 \\ 13.6 & 10.85 \end{bmatrix}$$

where the corresponding RGA is as

$$\Lambda_{p1}^{N}=\begin{bmatrix} 8.42 & -7.50 \\ -0.71 & 1.07 \\ -6.71 & 7.43 \end{bmatrix}$$

and

$$G_{p2}(0)=\begin{bmatrix} 9.77 & 7.53 \\ 13(1+0.1) & 10 \\ 13.6 & 10.85 \end{bmatrix}$$

where the corresponding RGA is as

$$\Lambda_{p2}^{N}=\begin{bmatrix} -1.24 & 1.47 \\ 8.44 & -7.47 \\ -6.20 & 6.99 \end{bmatrix}$$

comparison of Λ_{p1}^{N} and Λ_{p2}^{N} shows that the (2,1) element is more sensitive than the (1,1) element.

■

2.4.1.3 Squaring Down the Nonsquare Plant

Due to the difficulties associated with the direct analysis and control of nonsquare plants, squaring down the nonsquare multivariable plants is a common practice in practical control systems design. Especially in chemical process control, where it is generally known as the *selection of secondary measurements*. Consider the closed loop multivariable plant shown in figure 2.6, where $G(s)$ is a nonsquare plant and $G_c(s)$ is a square $m \times m$ controller matrix. The nonsquare plant matrix $G(s)$ is partitioned into a square matrix $G_S(s)$ and the complementary matrix $G_R(s)$. The subsequent plant equation is written as

$$\begin{bmatrix} Y_S(s) \\ Y_R(s) \end{bmatrix} = \begin{bmatrix} G_S(s) \\ G_R(s) \end{bmatrix} U(s) \tag{2.59}$$

where $Y_S(s)$ and $Y_R(s)$ are the m and $l-m$ dimensional selected controlled outputs and the remaining uncontrolled outputs, respectively. The main question is how to select these m outputs among the $l>m$ plant outputs? To answer this question, first note that there are $\dfrac{l!}{(l-m)!m!}$ possible combinations for $Y_S(s)$. The problem is now posed as: *Select the set of the controlled outputs such that the sum square steady state error of the remaining uncontrolled outputs is minimized, for any variation in the selected controlled outputs.*

Let us consider the controller design for the square part according to the IMC strategy shown in figure 2.5. Then, it can be easily verified that

$$U(s) = [I_m + G_c(s)(G_S(s) - \tilde{G}_S(s))]^{-1} G_c(s) R(s) \tag{2.60}$$

where $R(s)$ is the $m \times 1$ reference input for the $m \times 1$ controlled output $Y_S(s)$. Considering the steady state performance, we require

$$G_c(0) = \tilde{G}_S^{-1}(0) \tag{2.61}$$

For $\tilde{G}_S(0) = G_S(0)$, this gives

$$U(0) = \tilde{G}_S^{-1}(0) R \tag{2.62}$$

Also, in the case $\tilde{G}_S(0) \neq G_S(0)$, we have

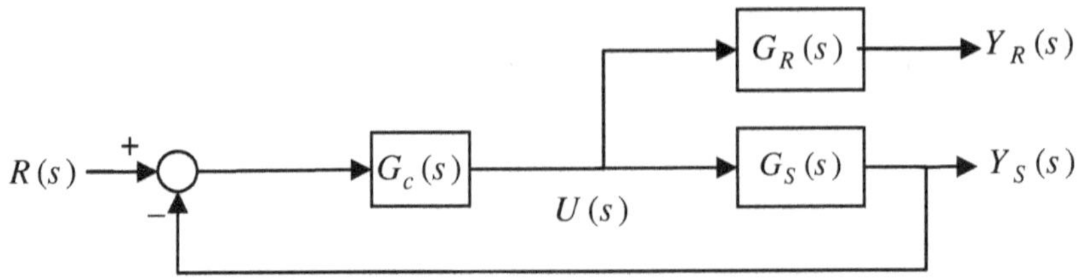

Fig. 2.6 Squaring down the nonsquare plant.

$$\begin{aligned} U(0) &= [I_m + \tilde{G}_S^{-1}(0)(G_S(0) - \tilde{G}_S(0))]^{-1}\tilde{G}_S^{-1}(0)R \\ &= [\tilde{G}_S(0)(I_m + \tilde{G}_S^{-1}(0)(G_S(0) - \tilde{G}_S(0)))]^{-1}R \\ &= G_S^{-1}(0)R \end{aligned} \tag{2.63}$$

Hence, the closed loop gains for the square plant is as equation (2.63). Assuming that the square part is under tight control, the steady state $l \times 1$ error vector for all the outputs is

$$E(0) = \left(\begin{bmatrix} I_m \\ 0 \end{bmatrix} - \begin{bmatrix} G_S(0) \\ G_R(0) \end{bmatrix} G_S^{-1}(0) \right) R = \left(I_{l \times m} - G(0) G_S^{-1}(0) \right) R \tag{2.64}$$

The sum squared error (SSE) for a given G_S is

$$\mathrm{SSE} = \sum_{i=1}^{m} \left\| E(i) \right\|_2^2 = \sum_{i=1}^{m} \left\| \left(I_{l \times m} - G(0) G_S^{-1}(0) \right) R_i \right\|_2^2 \tag{2.65}$$

where R_i is an $m \times 1$ vector with unity in the ith entry and zero elsewhere, and $E(i)$ is an $m \times 1$ error vector corresponding to R_i. The SSE problem given by (2.65) can be solved using the NSRGA. In fact, it is shown that the row sum of the NSRGA provides an exact optimal solution in two special cases and suboptimal solutions in other cases (Chang and YU 1990). The cases with optimal solutions are: $m = 1$ (SIMO plants), and $l - m = 1$.

For SIMO plants, using property 6 and equation (2.65) it is shown that

$$\mathrm{SSE}(i) = \frac{1}{rs(i)} - 1 \tag{2.66}$$

where $rs(i)$ is the ith row sum of the NSRGA. Therefore, to get the smallest SSE, the largest row sum of the NSRGA must be selected. On the other hand, property 6 indicates that the largest g_{i1} would gives the largest λ_{i1}^N and in turn the largest $rs(i)$. This is physically sensible. Since, common engineering sense would indicate the selection of the element with the most effect of manipulated input on the output.

For the case of $l-m=1$, there are $m+1$ choices to select. It is shown in (Chang and YU 1990) that the row sum of the NSRGA and the SSE for the ith selection in the $m+1$ possible choices follow the exact equation given by

$$\mathrm{SSE}(i)=\frac{rs(m+2-i)}{1-rs(m+2-i)} \tag{2.67}$$

Note that the value of row sum is between zero and one. Equation (2.67) clearly indicates that a small row sum gives a small SSE in the corresponding selected square plant matrix. Hence, to select a square subsystem, we eliminate the output with the *smallest* row sum in the NSRGA.

Unfortunately, an exact analytical relationship between the SSE and the NSRGA is not available for other cases. For multivariable 2-input plants a relationship is derived in (Chang and YU 1990). However, the heuristic of the above optimal solutions can be used as a starting point for the squaring down process in more general plants and experiments show that the NSRGA inference can result in the optimal or suboptimal selections in many cases.

Example 2.4.4

Consider the plant given in example 2.4.1. In this plant the condition of $l-m=4-3=1$ is satisfied. It is obvious that the third row sum of Λ^N is minimum and so we can eliminate the 3rd output corresponding to the minimum row sum in NSRGA. Therefore, eliminating the third output, the rest of the NSRGA is as follows

$$\Lambda^N = \begin{matrix} \\ y_1 \\ y_2 \\ y_4 \end{matrix}\begin{matrix} u_1 & u_2 & u_3 \\ \begin{bmatrix} 0.241 & -0.103 & 0.861 \\ 0.006 & 1.094 & -0.100 \\ 0.726 & 0.008 & 0.237 \end{bmatrix} \end{matrix}$$

and it proposes the $(u_1-y_4, u_2-y_2, u_3-y_1)$ as the appropriate input-output pair.

■

2.4.2 *Control Configuration Selection of Unstable Multivariable Plants*

The RGA and NI employed for the control configuration selection and integrity study in section 2.2.3, assumed open loop stability of the plant. This is a prime assumption in the preceding sections. The derived theories are widely based on the steady state conditions. However, steady state conditions are non existent for unstable plants and such assumptions are therefore not valid. In this section, these tools are extended to open loop unstable plants. To avoid the steady state term, zero frequency condition, i.e. $s = 0$, is used.

Consider the multivariable integral control shown in figure 2.1. The compensator matrix $C(s)$ is assumed diagonal and the open loop plant $G(s)$ is $m \times m$ and may be unstable with p poles in the RHP excluding the poles at the origin. Let

$$G_d = diag\{g_{ii}\} = \begin{bmatrix} g_{11} & & & 0 \\ & g_{22} & & \\ & & \ddots & \\ 0 & & & g_{mm} \end{bmatrix} \tag{2.68}$$

and p_d be the number of RHP poles of G_d. Also, define

$$G'_{ii} = \begin{bmatrix} g_{ii} & 0 \\ 0 & G^{ii} \end{bmatrix} \tag{2.69}$$

with p'_{ii} the number of its RHP poles.

The two following theorems from (Hovd and Skogestad 1994) are the main results available for the input-output pairing of unstable plants. The first theorem extends the NI, and the second theorem extends the RGA to unstable plants. The proofs are omitted and the interested reader should refer to (Hovd and Skogestad 1994).

Theorem 2.9
Consider the multivariable integral control of figure 2.1 with the aforementioned assumptions. Also, assume that $G(s)C(s)$ is strictly proper. Then, if the NI is defined as in equation (2.42) and

$$\text{NI}(G(0)) \begin{cases} <0 & \text{for} \quad p_d - p \quad \text{even} \\ >0 & \text{for} \quad p_d - p \quad \text{odd} \end{cases}$$

Al least one of the following instabilities will occur:
- The closed loop plant is unstable.
- At least one of the loops in the decentralized control is unstable.

■

Theorem 2.10
Consider the multivariable integral control of figure 2.1 with the aforementioned assumptions. Also, assume that $G(s)C(s)$ is strictly proper. Then, if the diagonal elements of the RGA are

$$\lambda_{ii}(0) \begin{cases} <0 & \text{for} \quad p'_{ii} - p \quad \text{even} \\ >0 & \text{for} \quad p'_{ii} - p \quad \text{odd} \end{cases}$$

At least one of the following instabilities will occur:

- The closed loop plant is unstable.
- The ith loop is unstable by itself.
- The plant is unstable as the ith loop is removed.

■

Note. The worst case in the above theorems is the closed loop instability. Also, the third case in theorem is detrimental, as it corresponds to situations such as actuator saturation which is common in practice.

Example 2.4.5

Consider the following transfer function matrix

$$G(s)=\begin{bmatrix}\dfrac{9s+1}{(-s+1)(s+1)} & \dfrac{2s-18}{(-s+1)(s+1)}\\ \dfrac{-1.5s-6}{(-s+1)(0.5s+1)} & \dfrac{12}{(-s+1)(0.5s+1)}\end{bmatrix}$$

To obtain the exact number of unstable poles of the multivariable plant, we evaluate its Smith-McMillan form (Maciejowski 1988). This gives

$$M(s)=\begin{bmatrix}\dfrac{1}{(-s+1)(s+1)(0.5s+1)} & 0\\ 0 & s+32\end{bmatrix}$$

Hence, the plant has only one unstable pole at 1 and $p=1$. On the other hand, G_d has two unstable poles at one and $p_d=2$. Therefore, $p_d-p=1$ is odd. Also, in this case, $G'_{11}=G_d$ and G'_{22} have the same number of unstable poles as G_d. Calculating the RGA and the NI for the plant, we have

$$\Lambda=\begin{bmatrix}-0.125 & 1.125\\ 1.125 & -0.125\end{bmatrix},\qquad \mathrm{NI}=-8$$

Therefore, theorems 2.9 and 2.10, based on stability considerations suggest the pairing (u_1-y_1,u_2-y_2). Note that one might wrongly choose the opposite pairing, based on the pairing rules for stable plants.

■

2.5 Conclusion

The main purpose of this chapter has been to provide an overview of the RGA approach to control configuration selection in linear multivariable plants.

The RGA was introduced as an interaction measure from the first principles under the assumption of tight control. Ten key RGA properties were introduced. These properties were derived from basic algebraic properties of the RGA definition and their implications were discussed. Then, the integrity properties of the feedback multivariable controllers with build in integral action were analyzed. The important tool of Niederlinski and relevant definitions for integrity analysis was introduced. The important notion of sensor or actuator failure was also dealt with. Then, the key idea of decentralized integral controllability was discussed. Based on these results, the pairing rules with three worked examples were given.

Finally, the RGA definition and properties which was originally derived for square stable multivariable plants were extended to nonsquare and unstable plants.

Extension to nonsquare plants was achieved by defining the NSRGA. Also, a practical way for squaring down of nonsquare plants was introduced using the NSRGA. Two theorems were given for the extension of NI and RGA to open loop unstable plants.

References

Bristol, E.H.: On a new measure of interaction for multivariable process control. IEEE Trans. Autom. Control 11, 133–134 (1966)

Campo, P.J., Morari, M.: Achievable closed loop properties of systems under decentralized control: Conditions involving the steady state gain. IEEE Trans. Autom. Control 39, 932–943 (1994)

Chang, J.W., Yu, C.C.: The relative gain for non-square multivariable systems. Chem. Eng. Sci. 45, 1309–1323 (1990)

Chiu, M.S., Arkun, Y.: Decentralized control structure selection based on integrity considerations. Ind. Eng. Chem. Res. 29, 369–373 (1990)

Grosdidier, P., Morari, M., Holt, B.R.: Closed-loop properties from steady state gain information. Ind. Eng. Chem. Fundam. 24, 221–235 (1985)

Hovd, M., Skogestad, S.: Simple frequency-dependent tools for control system analysis, structure selection and design. Automatica 28, 989–996 (1992)

Hovd, M., Skogestad, S.: Pairing criteria for decentralized control of unstable plants. Ind. Eng. Chem. Res. 33, 2134–2139 (1994)

Johansson, K.H.: The Quadruple-tank process: a multivariable laboratory process with an adjustable zero. IEEE Trans. Control Syst. Technol. 8, 456–465 (2000)

Kariwala, V., Hovd, M.: Relative Gain Array: Common misconceptions and clarifications. In: Proceedings of the 7th Symposium on Computer Process Control, Lake Louise, Canada (2006)

Kinnaert, M.: Interaction measures and pairing of controlled and manipulated variables for multiple-input-multiple-output systems: a survey. Journal A 36, 15–23 (1995)

Mc Avoy, T., Arkun, Y., Chen, R., Robinson, D., Schnelle, P.D.: A new approach to defining a dynamic relative gain. Control Eng. Pract. 11, 907–914 (2003)

Maciejowski, J.M.: Multivariable feedback design. Addison-Wesley, Reading (1989)

Morari, M., Zafiriou, E.: Robust process control. Prentice-Hall, New Jersey (1989)

Niederlinski, A.: A heuristic approach to the design of linear multivariable interacting control systems. Automatica 7, 691–701 (1971)

Porter, B., Khaki-Sedigh, A.: Robustness properties of tunable digital set-point tracking PID controllers for linear multivariable plants. Int. J. Control 49, 777–789 (1989)

Reeves, D.E., Arkun, Y.: Interaction measure for nonsquare decentralized control structures. AIChE Journal 35, 603–613 (1989)

Skogestad, S., Hovd, M.: Use of frequency-dependent RGA for control structure selection. In: Proceeding of the American Control Conference, pp. 2133–2139 (1990)

Skogestad, S., Morari, M.: Implications of large RGA elements on control performance. Ind. Eng. Chem. Res. 26, 2323–2330 (1987)

Skogestad, S., Morari, M.: Variable selection for decentralized control. Modeling, Identification, and Control 13, 113–125 (1992)

Skogestad, S., Postlethwaite, I.: Multivariable feedback control analysis and design. Wiley, Chichester (2005)

Treiber, S., Hoffman, D.W.: Multivariable constraint control using a frequency domain design approach. In: Proceeding of the 3rd conference on Chem. Proc. Control, Amsterdam (1986)

Wood, R.K., Berry, M.W.: Terminal composition control of a binary distillation column. Chem. Eng. Sci. 28, 1707–1717 (1973)

Chapter 3
Control Configuration of Linear Multivariable Plants: Advanced RGA Based Techniques

3.1 Introduction

The RGA introduced in chapter 2 provides a powerful tool for measuring control loop interaction and it is a well established pairing technique in the industry, with decades of successful applications. Although, the presented RGA analysis is sufficient for many practical problems, it is for some cases necessary to extend the RGA concept to handle certain shortcomings. This chapter provides an overview of the advanced pairing techniques based on the different RGA extensions. The presented methodologies are:

- **The Dynamic Relative Gain Array.** Dynamic relative gain is discussed in section 3.2.1. This is a dynamic extension of the RGA to improve the pairing capabilities of the steady state RGA in the cases where the RGA changes substantially with frequency and especially when the steady state RGA differs from the RGA at other key frequencies. In section 3.2.2, among the many different approaches, the *static output feedback linear quadratic regulator problem* strategy is chosen to develop the RGA dynamic extension. The Effective Relative Gain Array is another dynamic extension of the RGA presented in section 3.2.3, where the steady state gain and the response speed or plant bandwidth of the open loop transfer function are considered. Its key difference with the other RGA dynamic extensions is that no controller design or complex computations are necessary.
- **The Partial Relative Gain.** The RGA in cases where selected loops are closed in a control system can lead to incorrect pairing selections. In fact, partially controlled plants can improve the designer's knowledge of the plant under control. By investigating the relative gains of the uncontrolled part of the plant or the PRG, better pairing selections can be made. The notion of PRG is developed in section 3.3.
- **The Relative Interaction Array.** Section 3.4 introduces the relative interaction array (RIA). It is similar to the RGA and is defined based on individual control loops and it can only measure interaction in individual loops. The interaction measure provided by the RGA and the RIA can lead to several pairs that

A. Khaki-Sedigh and B. Moaveni: Control Configuration Selection, LNCIS 391, pp. 57–98.
springerlink.com

may satisfy the pairing rules and are unable to propose the best choice. However, an *overall interaction measure* is defined along with the RIA which claims to lead to the best input-output pairing selection.

- **Block Pairing and Block Relative Gain.** Fully decentralized control of *highly* interactive multivariable plants can result in poor closed loop performance. On the other hand, centralized controllers are not operationally desirable in certain complex multivariable plants where interaction is not heavily distributed through the plant and it is sever in some parts and less disturbing in the others. An effective design methodology in such cases is the *block decentralized controllers*. Block pairing discussed in section 3.5 is the first step in a successful block decentralized control system design.

3.2 The Dynamic Relative Gain Array

The steady state RGA analysis partly owes its widespread use to the simple open loop step response tests and straightforward computations required for the RGA calculation in real complex multivariable plants. However, the steady state RGA can result in erroneous pairing assessments in the following cases:

- The RGA changes substantially with frequency.
- The RGA near the *natural frequency* of a loop differs substantially from the steady state RGA.

In this section, we consider dynamic extensions of the RGA to improve the pairing capabilities of the steady state RGA. First, motivating examples are provided in section 3.2.1 to show the steady state RGA deficiencies. Then, a brief review on the literature is given and a basic dynamic extension of the RGA is defined. Sections 3.2.2 and 3.2.3 present more advanced dynamic extensions of the RGA.

3.2.1 Problem Formulation and a Basic Dynamic Extension

Two motivating examples from (McAvoy 1983) are presented to get a better insight into the above mentioned potential difficulties with the steady state RGA.

Example 3.2.1

The transfer function matrix of a two inputs and two outputs head box control system is derived in (McAvoy 1983) and is represented as follows

$$\begin{bmatrix} y_1(s) \\ y_2(s) \end{bmatrix} = \frac{1}{s+de} \begin{bmatrix} as & eb \\ -ad & b \end{bmatrix} \begin{bmatrix} u_1(s) \\ u_2(s) \end{bmatrix}$$

where a, b, d, and e are positive constants. It can be easily shown that

$$\lambda_{11}(s) = \frac{s}{s + ed}$$

Hence, we have the following two distinct cases:

1. $\lim_{s \to 0} \lambda_{11}(s) = 0$, which proposes the pairing $(u_1 - y_2, u_2 - y_1)$.
2. $\lim_{s \to \infty} \lambda_{11}(s) = 1$, which proposes the pairing $(u_1 - y_1, u_2 - y_2)$.

It is claimed in (McAvoy 1983) that the pairing $(u_1 - y_1, u_2 - y_2)$ is used in actual practice, which contradicts the conventional steady state RGA pairing. In fact, the effect of u_1 on y_1 is much faster than that of u_2. Also, both inputs affect y_2 slowly. Hence, the loop $u_1 - y_1$ will be fast and can be tuned tightly.

■

Example 3.2.2

Consider the two transfer function matrices as follows

$$G_1(s) = \begin{bmatrix} \dfrac{0.35}{(7.74s+1)^2} & \dfrac{-0.516e^{-0.5s}}{(7.1s+1)^2} \\ \dfrac{0.5e^{-0.5s}}{(15.8s+1)(0.5s+1)} & \dfrac{-0.9}{(13.8s+1)(0.4s+1)} \end{bmatrix}$$

and

$$G_2(s) = \begin{bmatrix} \dfrac{-0.2}{(18.3s+1)(5.6s+1)} & \dfrac{0.1}{(5.76s+1)(1.25s+1)} \\ \dfrac{-0.4e^{-0.3s}}{(28.3s+1)(0.62s+1)} & \dfrac{-0.05}{(3.3s+1)} \end{bmatrix}$$

Calculating the dynamic RGA, $\Lambda(j\omega)$ and plotting $\lambda_{11}(j\omega)$ for a given frequency range, the frequency response results of figure 3.1 are obtained. It is observed from figure 3.1 that the magnitude of $\lambda_{11}(j\omega)$ corresponding to $G_1(s)$ has considerable changes with frequency yet results in the diagonal pairing for all frequencies. Since, it is always greater than 0.5. On the other hand, the magnitude of the RGA for $G_2(s)$ proposes the off-diagonal pairing for low frequencies (it is less than 0.5) and diagonal pairing for high frequencies (it is greater than 0.5), although the variation of the corresponding RGA magnitude remains relatively unchanged as a function of frequency. If the working frequencies of the two control loops are known, the frequency results of figure 3.1 can be further used to study the closed loop performance of the control strategies. For instance, at lower frequencies the RGA values for $G_1(s)$ are high, which indicates performance degradation.

■

The pioneer work in defining the dynamic RGA (DRGA) based interaction measure is presented for a two inputs two outputs plant in (Witcher and McAvoy 1977), where a transfer function model of the plant is used instead of its steady state gain matrix in the definition of the RGA. This work was later generalized in a monograph (McAvoy 1983). It is obvious that the dynamic RGA would be more complex than the RGA, and we can loose some of the nice features of the RGA in any dynamic extension.

A key feature of the RGA for control configuration selection is its independence of the controller design, disturbances and minimal dependence on the plant models. These features are lost in the DRGA by including detailed controller design in some methods, such as (Bristol 1979) and (Tung and Edgar 1981). Another approach to define the DRGA can be found in (Gagnepain and Seborg 1982). Bristol defined the DRGA as

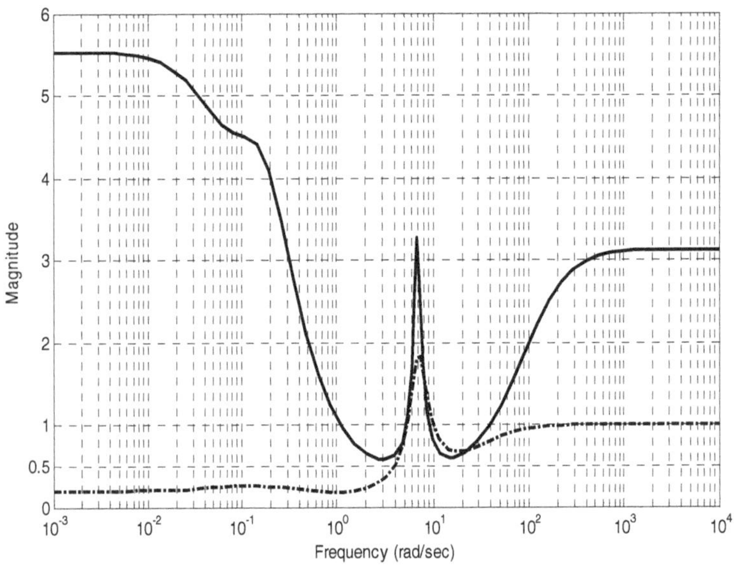

Fig. 3.1 Frequency response of the DRGA, Solid line corresponds to $\lambda_{11}(s)$ for $G_1(s)$; Dash-dot line corresponds to $\lambda_{11}(s)$ for $G_2(s)$.

$$\Lambda(s) = G(s).*G^{-T}(s) \tag{3.1}$$

By replacing $s = j\omega$ in equation (3.1), the elements of DRGA can be plotted versus frequency. The pairing should be such that, over the frequency range of interest, the magnitude of the diagonal elements is close to one and the magnitude of the off diagonal elements is small. This is valid for plants that are not ill conditioned, where significant one way interaction is available, as in the triangular plants. It can be shown that the algebraic properties of the DRGA and the RGA are similar.

In (Hovd and Skogestad 1992), the DRGA is used to derive bounds on the designs of the individual loops to ensure the desired closed loop performance for both set point tracking and disturbance rejection after closing all the loops. A generalized dynamic relative gain (GDRG) is defined in (Huang et al. 1994). This approach consists of two steps, where some infeasible pairings are rejected by the steady state RGA and then using the proposed method desirable pairs is determined.

In (McAvoy et al. 2003), a new approach to defining a DRGA is presented. This approach requires a state space dynamical model of the plant. If a transfer function model is given, its realization must be derived. A proportional output optimal feedback controller is designed and the resulting controller gain matrix is used to define the new DRGA.

In what follows we present two examples to show the concepts of the basic DRGA for 2×2 multivariable plants.

Example 3.2.3

Consider the following 2×2 multivariable plant

$$y_1(s) = g_{11}(s)u_1(s) + g_{12}(s)u_2(s)$$
$$y_2(s) = g_{21}(s)u_1(s) + g_{22}(s)u_2(s)$$

Using equation (3.1) and defining

$$\kappa = \frac{g_{12}(s)g_{21}(s)}{g_{11}(s)g_{22}(s)}$$

the DRGA is

$$\Lambda(s) = \begin{bmatrix} \dfrac{1}{1-\kappa} & -\dfrac{\kappa}{1-\kappa} \\ -\dfrac{\kappa}{1-\kappa} & \dfrac{1}{1-\kappa} \end{bmatrix}$$

Then, by setting $s = j\omega$ the frequency response analysis of the DRGA is carried out. The magnitude diagram can now be interpreted in exactly the same way as the steady state RGA. The magnitude of $\lambda_{11}(j\omega)$ is a measure of interaction at a particular frequency. A value of 1 corresponds to $\kappa = 0$, which is a decoupled situation. A value of 0.5 is indicative of high interactions.

■

Example 3.2.4

Consider the 2×2 multivariable plant in example 3.2.3, with a first order time delayed model as

$$g_{ij}(s) = \frac{K_{ij}e^{-\tau_{ij}s}}{T_{ij}s+1} \qquad i,j = 1,2$$

then

$$\kappa = \left(\frac{K_{12}K_{21}}{K_{11}K_{22}}\right)\left(\frac{T_{11}s+1}{T_{12}s+1}\right)\left(\frac{T_{22}s+1}{T_{21}s+1}\right)e^{-(\tau_{12}+\tau_{21}-\tau_{11}-\tau_{22})s}$$

We now study the effects of the various parameters in κ. As delay does not affect the magnitude term, it therefore does not show the interaction degree. Note that in the steady state, this equation easily leads to the well known result, $\lambda_{11} = 1 \Big/ \left(1 - \frac{K_{12}K_{21}}{K_{11}K_{22}}\right)$. However, the time constants are the *dynamic gains*. Significant time constant differences in the plant lead to substantial change in the magnitude of κ with frequency. Let (Witcher and McAvoy 1977)

$$g_{11}(s) = g_{22}(s) = \frac{2e^{-s}}{(10s+1)}$$

and

$$g_{12}(s) = g_{21}(s) = \frac{0.5e^{-s}}{(s+1)}$$

Figure 3.2 shows the magnitude of κ and $\lambda_{11}(s)$ as a function of frequency. It is observed that the steady state RGA is

$$\Lambda = \begin{bmatrix} 1.066 & -0.066 \\ -0.066 & 1.066 \end{bmatrix}$$

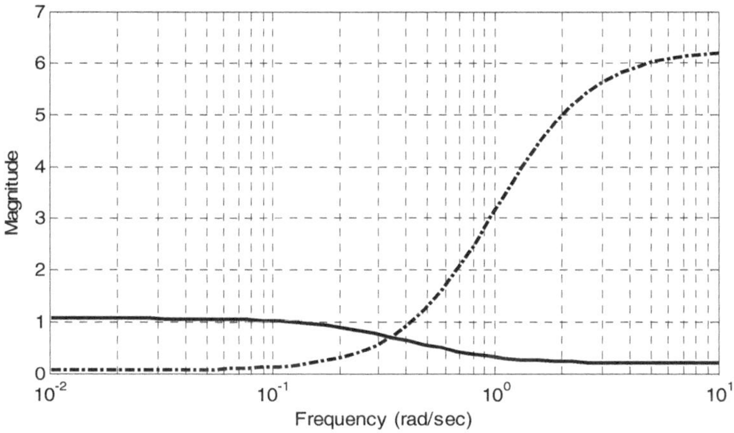

Fig. 3.2 Frequency response of the DRGA, Solid line corresponds to $\lambda_{11}(s)$ and Dash-dot line corresponds to κ.

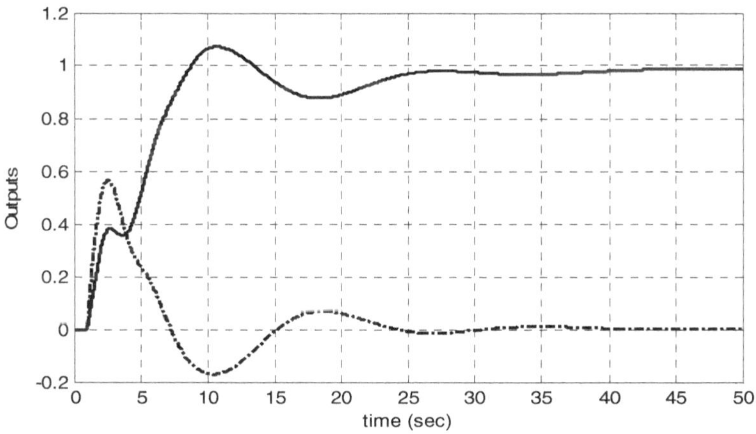

Fig. 3.3 Closed loop responses, Solid line shows the first output and Dash-dot line corresponds to the second output.

which indicates the $(u_1 - y_1, u_2 - y_2)$ pairing. However, figure 3.2 shows that such a pairing leads to a highly interactive plant at middle frequencies, and at high frequencies the appropriate input-output pair is $(u_1 - y_2, u_2 - y_1)$. Closed loop responses are now used to evaluate the control system performance. Figure 3.3 shows the closed loop step responses with the PI controllers for the diagonal pairing $G_{c1}(s) = G_{c2}(s) = 1.5(1 + 0.06/s)$, the set point for the first output is unit step and zero for the second output. The responses show the highly interactive transient responses of the plant.

■

Note. The DRGA can be evaluated using equation (3.1). However, in the cases where an adequately accurate model of the plant is not available, experimental data obtained from simple open loop responses of the plant can be used to calculate the DRGA. A procedure for such tests is given in (Witcher and McAvoy 1977) and (Shinskey 1967). Also, these results can be extended to the general $n \times n$ multivariable plants.

3.2.2 An Advanced Dynamic Extension of the RGA

In this section, following the results of (McAvoy et al. 2003), a new approach to defining a DRGA is presented. Let the *scaled* state space description of the multivariable plant be

$$\begin{aligned} \dot{x}(t) &= Ax(t) + Bu(t) \\ y(t) &= Cx(t) \end{aligned} \tag{3.2}$$

with $x(0) = x_0$, where $x(t) \in \Re^n$, $u(t)$ and $y(t) \in \Re^m$. The scaled state space model will indicate that the inputs are scaled by their operating range and the outputs are scaled by the range in which it is desired to hold them. The state scaling only affects the initial condition and the steady state values are used as the scaling factor.

Let the controller be static proportional output feedback given by

$$u(t) = -Ky(t) \tag{3.3}$$

where K is a constant square matrix. It is desired to apply the control law given by equation (3.3) to minimize the following cost function

$$J = \frac{1}{2}\int_0^{\infty}(y^T Q y + u^T R u)dt \tag{3.4}$$

where the designer selects the weighting matrices. This is a *static output feedback linear quadratic regulator problem.* Standard theories are developed to solve the given optimization problem. See for example (Levine and Athans 1970) or (Lewis 1992). Note that the optimization solution matrix K is calculated based on the state space model of the plant. It therefore contains ample information regarding the plant dynamics and can be used to derive the new definition of the DRGA.

A new definition of the DRGA is now presented. The $(i,j)^{th}$ element of the DRGA is defined as

$$\lambda_{Dij} \triangleq \frac{\partial u_j / \partial y_i \big|_{u_k \neq 0, k \neq j}}{\partial u_j / \partial y_i \big|_{u_k = 0, k \neq j}} \tag{3.5}$$

The numerator and denominator of equation (3.5) give the gain of u_j to y_i assuming that the plant is controlled with the control law given by equation (3.3) and the optimal output gain calculated from the optimization process of minimizing equation (3.4) subject to equations (3.2) and (3.3). The numerator gives the change in input u_j to a change in output y_i, for the optimal closed loop plant. That is, in the transient stage where the optimal controller is regulating the plant to the origin, numerator gives the change in u_j resulting from a change in y_i. For the denominator, only one input u_j changes. This is an idealized scenario where it is assumed that plant outputs are optimally driven to the origin from a random initial condition and only the jth input changes. Using equation (3.3), the partial derivatives in equation (3.5) can be calculated to give

$$\lambda_{Dij} = \frac{-K_{ij}}{-1/\widehat{K}_{ji}} = K_{ij}\widehat{K}_{ji} \tag{3.6}$$

where $\widehat{K} = K^{-1}$.

Example 3.2.5

Consider the following transfer function matrix

$$G(s) = \begin{bmatrix} \dfrac{5}{100s+1} & \dfrac{k_{12}}{10s+1} \\ \dfrac{k_{21}}{10s+1} & \dfrac{5}{100s+1} \end{bmatrix}$$

with values of $k_{12} = 1$, and $k_{21} = -5$, 1, and 5. Note that the off diagonal terms in the transfer function matrix have faster dynamics than the diagonal elements. To calculate the new DRGA, state space realization of the transfer function matrix for the 3 cases is derived.

Case 1: $k_{21} = 1$

The minimal state space realization is

$$\dot{x} = \begin{bmatrix} -0.11 & -0.001 & 0 & 0 \\ 1 & 0 & 0 & 0 \\ 0 & 0 & -0.11 & -0.001 \\ 0 & 0 & 1 & 0 \end{bmatrix} x + \begin{bmatrix} 1 & 0 \\ 0 & 0 \\ 0 & 1 \\ 0 & 0 \end{bmatrix} u$$

$$y = \begin{bmatrix} 0.05 & 0.005 & 0.1 & 0.001 \\ 0.1 & 0.001 & 0.05 & 0.005 \end{bmatrix} x$$

and the corresponding RGA is

$$\Lambda = \begin{bmatrix} 1.041 & -0.041 \\ -0.041 & 1.041 \end{bmatrix}$$

which proposes the diagonal pairing. By choosing $Q = C^T C$ and $R = I$ the proportional output feedback is

$$K = \begin{bmatrix} 0.2376 & 0.3938 \\ -0.0079 & 0.9128 \end{bmatrix}$$

and the DRGA is

$$\Lambda_D = K.*K^{-T} = \begin{bmatrix} 0.9858 & 0.0142 \\ 0.0142 & 0.9858 \end{bmatrix}$$

which similarly proposes the diagonal pair.

Case 2: $k_{21} = 5$
Calculating the proportional output feedback with the same Q and R matrices, gives

$$K = \begin{bmatrix} 0.2751 & 0.7728 \\ 0.4412 & 0.0559 \end{bmatrix}$$

and the corresponding DRGA is

$$\Lambda_D = \begin{bmatrix} -0.0473 & 1.0473 \\ 1.0473 & -0.0473 \end{bmatrix}$$

which recommends the off-diagonal pairing. However, the corresponding RGA is as follows and proposes the diagonal pairing.

$$\Lambda = \begin{bmatrix} 1.25 & -0.25 \\ -0.25 & 1.25 \end{bmatrix}$$

Case 3: $k_{21} = -5$

The proportional output feedback and the corresponding DRGA are

$$K = \begin{bmatrix} 0.1388 & -0.7983 \\ 0.7145 & 0.2149 \end{bmatrix}$$

and

$$\Lambda_D = \begin{bmatrix} 0.0497 & 0.9503 \\ 0.9503 & 0.0497 \end{bmatrix}$$

which recommends the off diagonal input-output pair. However, it can be show that the RGA proposes the diagonal pairing.

■

Example 3.2.6

Consider the transfer function matrix of example 2.3.3 as

$$G(s) = \begin{bmatrix} \dfrac{5e^{-40s}}{100s+1} & \dfrac{e^{-4s}}{10s+1} \\ \dfrac{-5e^{-4s}}{10s+1} & \dfrac{5e^{-40s}}{100s+1} \end{bmatrix}$$

where the corresponding RGA is given below and it recommends the diagonal pairing.

$$\Lambda = \begin{bmatrix} 0.8333 & 0.1667 \\ 0.1667 & 0.8333 \end{bmatrix}$$

Using a first order Pade approximation, the corresponding state space realization $S(A, B, C)$ is derived

$$A = \begin{bmatrix} -0.66 & -0.86 & -0.0033 & -2.5\times10^{-5} & 0 & 0 & 0 & 0 \\ 1 & 0 & 0 & 0 & 0 & 0 & 0 & 0 \\ 0 & 1 & 0 & 0 & 0 & 0 & 0 & 0 \\ 0 & 0 & 1 & 0 & 0 & 0 & 0 & 0 \\ 0 & 0 & 0 & 0 & -0.66 & -0.86 & -0.003 & -2.5\times10^{-5} \\ 0 & 0 & 0 & 0 & 1 & 0 & 0 & 0 \\ 0 & 0 & 0 & 0 & 0 & 1 & 0 & 0 \\ 0 & 0 & 0 & 0 & 0 & 0 & 1 & 0 \end{bmatrix}$$

$$B^T = \begin{bmatrix} 1 & 0 & 0 & 0 & 0 & 0 & 0 & 0 \\ 0 & 0 & 0 & 0 & 1 & 0 & 0 & 0 \end{bmatrix}$$

$$C = \begin{bmatrix} -0.05 & -0.0275 & -0.001 & 0.0001 & -0.1 & 0.044 & 0.003 & 2.5\times10^{-5} \\ -0.5 & 0.22 & 0.0147 & 0.0001 & -0.05 & -0.0275 & 0-0.001 & 0.0001 \end{bmatrix}$$

Then, the proportional output feedback for $Q = C^T C$ and $R = I$ is

$$K = \begin{bmatrix} -0.29 & -0.39 \\ 0.47 & -0.20 \end{bmatrix}$$

and the corresponding DRGA is

$$\Lambda_D = K.*K^{-T} = \begin{bmatrix} 0.25 & 0.75 \\ 0.75 & 0.25 \end{bmatrix}$$

It is obvious that the DRGA proposes the off-diagonal pair. In example 2.3.3, using two optimized PI controllers proposed by McAvoy (2003), the closed loop step responses for diagonal and off-diagonal pairing are analyzed and the comparison between the responses shows the superiority of diagonal pairing proposed by the RGA method.

■

3.2.3 *The Effective Relative Gain Array*

The DRGA introduced in section 3.2.1, overcomes the limitations of the static RGA by considering the effects of the process dynamics in the pairing process. Similar to the RGA, denominators of the DRGA elements are calculated under the assumption of perfect control at all frequencies, and the numerators are the open loop transfer functions. A common characteristic of nearly all the DRGA definitions is their controller dependence nature. The method developed in section 3.2.2

is based on an optimal static output feedback controller design. In a rather different approach to overcome the deficiencies of steady state RGA, the effective RGA (ERGA) is introduced in (Xiong et al. 2005). This is an extension of the RGA, where two factors in the open loop transfer function are considered: the steady state gain and the response speed or plant bandwidth. Hence, no controller design or complex computations are necessary. This section follows closely the results of (Xiong et al. 2005).

Consider a stable multivariable plant described by a $m \times m$ transfer function matrix $G(s)$ with nonsingular $G(0)$. A key observation made in defining the ERGA is the fact that perfect or tight control assumption is only valid for very low frequency range, while in the decentralized control system designs, the loops may be tuned around their critical frequency regions or the control system bandwidths. Hence, this frequency region is important in considering the interaction effects. Therefore, it is concluded in (Xiong et al. 2005) that the following two open loop characteristics influence the configuration selection:

- *Steady state gain* $g_{ij}(0)$ that reflects the effect of the input u_j on the output y_i.
- *Response speed* that accounts for the sensitivity of the output y_i to the input u_j. This is indicative of the interaction rejection ability of the plant.

Now let

$$g_{ij}(j\omega) = g_{ij}(0) g_{ij}^{0}(j\omega) \tag{3.7}$$

where $g_{ij}^{0}(j\omega)$ is the normalized transfer function of $g_{ij}(j\omega)$, i.e. $g_{ij}^{0}(0) = 1$, and define the *effective gain* e_{ij} for a particular transfer function as

$$e_{ij} = g_{ij}(0) \int_{0}^{\omega_{C,ij}} \left| g_{ij}^{0}(j\omega) \right| d\omega \tag{3.8}$$

where $\omega_{C,ij}$ for $i, j = 1, 2, \cdots, m$ are the $g_{ij}^{0}(j\omega)$ *critical frequencies* and $|\bullet|$ denotes the absolute value. The critical frequency is defined in two ways in (Xiong and Cai 2006). It can be taken equal to $\omega_{B,ij}$ for $i, j = 1, 2, \cdots, m$, where $\omega_{B,ij}$ is the *bandwidth* of $g_{ij}^{0}(j\omega)$. The bandwidths are the frequencies from which the zero frequency or the steady state gain is attenuated more than 3dB. That is, the magnitude of the frequency response drops below $0.707 g_{ij}(0)$, i.e. $g_{ij}(j\omega_{B,ij}) = 0.707 g_{ij}(0)$. This definition is employed for transfer function matrices with some elements without phase crossover frequencies, such as first order or second order without time delay. The other definition for critical frequency is $\omega_{U,ij}$ for $i, j = 1, 2, \cdots, m$, where $\omega_{U,ij}$ is the *ultimate frequency* of $g_{ij}^{0}(j\omega)$. The ultimate frequency is the frequency where the phase plot crosses $-\pi$, i.e. $\arg\left[g_{ij}^{0}(j\omega_{U,ij})\right] = -\pi$. Without loss of generality, in what follows we will use $\omega_{B,ij}$ as the critical frequency.

Effective gain calculation. Consider the frequency response curve of $g_{ij}(j\omega)$ shown in figure 3.4. Equation (3.8) is equal to the area under the curve from 0 up to the $\omega_{B,ij}$ frequency. The integral can be approximated by a rectangular area as follows

$$e_{ij} \approx g_{ij}(0)\omega_{B,ij} \qquad i, j = 1, 2, \cdots, m \tag{3.9}$$

Also, note that $\left|g_{ij}^{0}(j\omega)\right|$ represents the magnitude of the normalized transfer function at each frequency. Hence, the summation of these magnitudes can be considered as the *effective energy output* of $g_{ij}(j\omega)$. The following definition now follows.

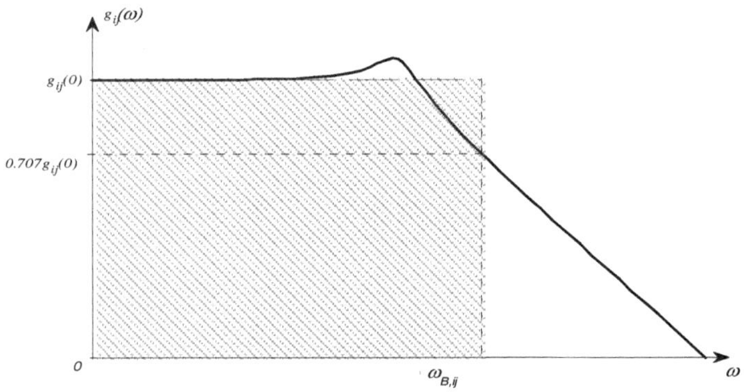

Fig. 3.4 Frequency response curve of $g_{ij}(j\omega)$ and the corresponding effective gain.

Definition 3.1 The *effective gain matrix* is defined as

$$E = \begin{bmatrix} e_{11} & e_{12} & \cdots & e_{1m} \\ e_{21} & e_{22} & \cdots & e_{2m} \\ \cdots & \cdots & \cdots & \cdots \\ e_{m1} & e_{m2} & \cdots & e_{mm} \end{bmatrix} \tag{3.10}$$

which using equation (3.9), can be written as

$$\mathrm{E} = G(0).*\,\Omega \tag{3.11}$$

where $G(0)$ is the steady state gain matrix of the plant and Ω is the *bandwidth matrix* defined as

$$\Omega = \begin{bmatrix} \omega_{B,11} & \omega_{B,12} & \cdots & \omega_{B,1m} \\ \omega_{B,21} & \omega_{B,22} & \cdots & \omega_{B,2m} \\ \cdots & \cdots & \cdots & \cdots \\ \omega_{B,m1} & \omega_{B,m2} & \cdots & \omega_{B,mm} \end{bmatrix} \tag{3.12}$$

Note that when the output y_i is under control using the input u_j, e_{ij} is an indication of interaction energy to the other loops. In fact, the value of e_{ij}, relates to the dominance of the loop. Bigger values of the e_{ij} will give more dominant loops. This leads to the next definition.

Definition 3.2 The *effective relative gain* between y_i and u_j is defined as

$$\varepsilon_{ij} = \frac{e_{ij}}{\hat{e}_{ij}} \tag{3.13}$$

where $\hat{e}_{ij}$ is the effective gain between y_i and u_j when all other loops are closed. Then, the *effective relative gain array* (ERGA) is defined as

$$\Gamma_{ERGA} = \begin{bmatrix} \varepsilon_{11} & \varepsilon_{12} & \cdots & \varepsilon_{1m} \\ \varepsilon_{21} & \varepsilon_{22} & \cdots & \varepsilon_{2m} \\ \cdots & \cdots & \cdots & \cdots \\ \varepsilon_{m1} & \varepsilon_{m2} & \cdots & \varepsilon_{mm} \end{bmatrix}$$

and it can be calculated as

$$\Gamma_{ERGA} = \mathrm{E}.*\,\mathrm{E}^{-T} \tag{3.14}$$

where "$.*$" is the element by element product and E is the effective gain matrix defined by equation (3.10).

Note Both RGA and ERGA use relative gains, therefore the RGA properties can be extended to ERGA.

The ERGA based pairing rules. The following pairing rules are proposed using the ERGA. The arguments are similar to that of section 2.3 for the RGA. The pairing rules can be summarized as:

- Choose the input-output pair with ERGA elements close to 1.
- Niederlinski Index must be positive.
- ERGA elements corresponding to the input-output pair must be positive.
- Large ERGA elements are not appropriate for input-output pairing.

Note 1. Effective gains in ERGA replace the steady state information used in the RGA. Hence, dynamic interaction up to the bandwidth is considered in the pairing.

Note 2. ERGA uses both the steady state gains and the bandwidth information. It is thus more computationally involved than the RGA but easier to calculate than the DRGA.

Note 3. ERGA uses open loop information *only* and is therefore controller independent.

Example 3.2.7

Consider a 2×2 multivariable plant given by the following transfer function matrix

$$G(s)=\begin{bmatrix} \dfrac{5}{4s+1} & \dfrac{0.5e^{-5s}}{(2s+1)(15s+1)} \\ \dfrac{-4e^{-6s}}{20s+1} & \dfrac{1}{3s+1} \end{bmatrix}$$

The corresponding RGA is

$$\Lambda=\begin{bmatrix} 0.7143 & 0.2857 \\ 0.2857 & 0.7143 \end{bmatrix}$$

and diagonal pairing is therefore proposed. The Bandwidth matrix for the plant is calculated as

$$\Omega = \begin{bmatrix} 0.2494 & 0.0654 \\ 0.0499 & 0.3325 \end{bmatrix}$$

Then, using equation (3.11) we have

$$\mathrm{E} = \begin{bmatrix} 5 & 0.5 \\ -4 & 1 \end{bmatrix} .* \begin{bmatrix} 0.2494 & 0.0654 \\ 0.0499 & 0.3325 \end{bmatrix} = \begin{bmatrix} 1.2470 & 0.0327 \\ -0.1995 & 0.3325 \end{bmatrix}$$

and using equation (3.14) the ERGA is

$$\Gamma_{ERGA} = \begin{bmatrix} 0.9845 & 0.0155 \\ 0.0155 & 0.9845 \end{bmatrix}$$

Thus, the ERGA proposes the diagonal pairing, which agrees with the RGA result.

■

Example 3.2.8

Consider the following multivariable plant (Huang et al. 1994)

$$G(s) = \begin{bmatrix} \dfrac{-2e^{-s}}{10s+1} & \dfrac{1.5e^{-s}}{s+1} & \dfrac{e^{-s}}{s+1} \\ \dfrac{1.5e^{-s}}{s+1} & \dfrac{e^{-s}}{s+1} & \dfrac{-2e^{-s}}{10s+1} \\ \dfrac{e^{-s}}{s+1} & \dfrac{-2e^{-s}}{10s+1} & \dfrac{1.5e^{-s}}{s+1} \end{bmatrix}$$

The corresponding RGA is

$$\Lambda = \begin{bmatrix} -0.9302 & 1.1860 & 0.7442 \\ 1.1860 & 0.7442 & -0.9302 \\ 0.7442 & -0.9302 & 1.1860 \end{bmatrix}$$

The RGA proposes the pairings $(u_2 - y_1, u_1 - y_2, u_3 - y_3)$ and $(u_3 - y_1, u_2 - y_2, u_1 - y_3)$. The two selected pairings have similar RGA elements properties and are indifferent from the RGA viewpoint. To use the ERGA, the bandwidth matrix of the plant is calculated as

$$\Omega = \begin{bmatrix} 0.1 & 1 & 1 \\ 1 & 1 & 0.1 \\ 1 & 0.1 & 1 \end{bmatrix}$$

and the corresponding ERGA is

$$\Gamma_{ERGA} = \begin{bmatrix} 0.0554 & 0.6977 & 0.2468 \\ 0.6977 & 0.2468 & 0.0554 \\ 0.2468 & 0.0554 & 0.6977 \end{bmatrix}$$

The ERGA proposes the pairing $(u_2 - y_1, u_1 - y_2, u_3 - y_3)$, as the corresponding ERGA elements are closer to one. This is also validated by the result of (Huang et al. 1994).

■

Example 3.2.9

Consider the following 2×2 multivariable plant

$$G(s) = \begin{bmatrix} \dfrac{1}{(s+1)} & \dfrac{-k}{(s+4)} \\ \dfrac{k}{(s+4)} & \dfrac{1}{(s+1)} \end{bmatrix}$$

We will discuss the pairing problem for different values of $k > 0$. It can be easily shown that the NI and the RGA elements are positive for all k. Also, the diagonal elements and the absolute values of the off-diagonal elements are equal. Consider the following two cases:

Case 1: Let $k = 4$, then

$$G(s) = \begin{bmatrix} \dfrac{1}{(s+1)} & \dfrac{-4}{(s+4)} \\ \dfrac{4}{(s+4)} & \dfrac{1}{(s+1)} \end{bmatrix}$$

and

$$\Lambda(G) = \begin{bmatrix} 0.5 & 0.5 \\ 0.5 & 0.5 \end{bmatrix}$$

The static RGA is unable to distinguish between the two pairings. The frequency responses of the dynamic RGA in case 1 are shown in figure 3.5.

Figure 3.5 shows that the off-diagonal pairing is the dominant pair in all frequencies. The absolute value of zero frequency gain for all the transfer function elements is the same. So there is no pairing priority from the gain point of view. But, the off diagonal elements have wider bandwidth compared with the diagonal ones. Therefore, the controller affect on these elements takes place in a wider range of frequency. In addition, the diagonal elements vanish at an earlier range of frequencies. Thus, without any calculation, we can claim that the off diagonal pairing is preferred in this case. This result can clearly be seen by the ERGA matrix:

$$\Gamma_{ERGA} = \begin{bmatrix} 0.06 & 0.94 \\ 0.94 & 0.06 \end{bmatrix}$$

Case 2: Let $k = 2$, then

$$G(s) = \begin{bmatrix} \dfrac{1}{(s+1)} & \dfrac{-2}{(s+4)} \\ \dfrac{2}{(s+4)} & \dfrac{1}{(s+1)} \end{bmatrix}$$

and

$$\Lambda = \begin{bmatrix} 0.8 & 0.2 \\ 0.2 & 0.8 \end{bmatrix}$$

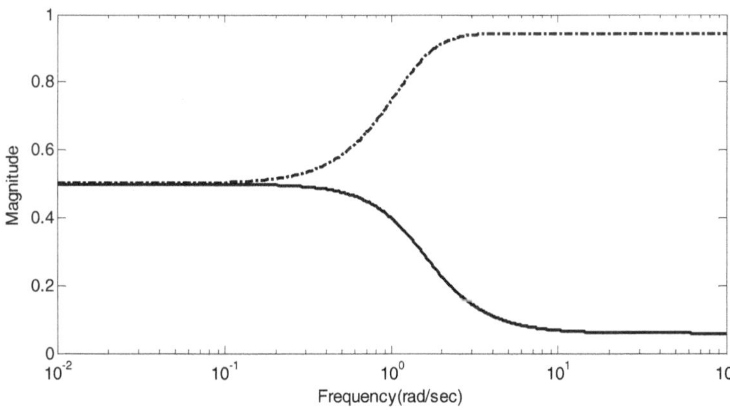

Fig. 3.5. The frequency responses of dynamic RGA for $k = 4$. Solid line corresponds to $\lambda_{11}(s)$ and Dash-dot line corresponds to $\lambda_{12}(s)$.

Hence, the diagonal pairing is proposed by the RGA. Calculating the ERGA gives

$$\Gamma_{ERGA} = \begin{bmatrix} 0.2 & 0.8 \\ 0.8 & 0.2 \end{bmatrix}$$

which proposes the off diagonal pairing. On the other hand, the frequency response of the dynamic RGA is shown in figure 3.6.

As it can be seen from figure 3.6, the diagonal pairing is recommended in lower range of frequencies, while the off diagonal pairing is preferred in higher frequencies. Therefore, the best pairing depends on the desired closed loop bandwidth of the plant.

■

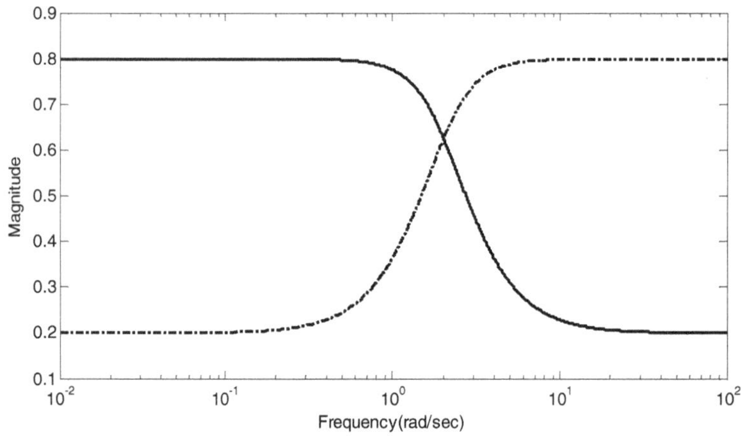

Fig. 3.6 The frequency responses of dynamic RGA for $k = 2$. Solid line corresponds to $\lambda_{11}(s)$ and Dash-dot line corresponds to $\lambda_{12}(s)$.

3.3 The Partial Relative Gain

The RGA was defined in section 2.2.1 and its properties were developed in sections 2.2.2 and 2.2.3. Using the derived properties the conventional pairing rules are given in section 2.3. According to these pairing rules, a pairing with $\lambda_{ij} = 0$ should not be selected. However, it is a well known fact that such a pairing may in certain cases result in a stable and well behaved closed loop plant. The key point is that the pairing would very much depend on the design of the other control loops. This issue is addressed in (Haggblom 1997a,b) and leads to the so called Partial Relative Gain (PRG). In the following examples from (Haggblom 1997b), it is shown that in certain cases, the closure of control loops may provide useful information about the control configuration. This type of information may not be attainable from simple open loop tests as in the classical RGA.

Example 3.3.1

The open loop transfer function matrix of a two-product distillation column with a total condenser can be written as

$$\begin{bmatrix} y(s) \\ z(s) \end{bmatrix} = \begin{bmatrix} G_{yu}(s) & 0 \\ G_{zu}(s) & Is^{-1} \end{bmatrix} \begin{bmatrix} u(s) \\ v(s) \end{bmatrix}$$

where $y(s)$ and $z(s)$ are the vectors of product qualities and column inventories, respectively. Also, $u(s)$ and $v(s)$ are the vectors of internal flow rates and product flow rates, respectively. The RGA for this plant is

$$\Lambda(G) = \begin{bmatrix} G_{yu}(s) & 0 \\ G_{zu}(s) & Is^{-1} \end{bmatrix} \cdot * \begin{bmatrix} G_{yu}(s) & 0 \\ G_{zu}(s) & Is^{-1} \end{bmatrix}^{-T}$$

where "$.*$" is the element by element product. This leads to

$$\Lambda(G) = \begin{bmatrix} \Lambda_{yu}(s) & 0 \\ 0 & I \end{bmatrix}$$

where $\Lambda_{yu}(s) = G_{yu}(s) \cdot * G_{yu}(s)^{-T}$ is the RGA for $G_{yu}(s)$. The pairing rules of section 2.3 would *only* recommend the pairing $(u - y, v - z)$ for the two-product distillation column. Note that any other pairing would require a pairing corresponding to a zero relative gain. This control configuration is commonly called the LV-structure. However, in the practicing control engineering methodologies the so called DB-structure is well established, where the recommended pairing is $(u - z, v - y)$. This configuration becomes feasible when the inventory control loops are closed. Hence, to evaluate such control configurations, it is necessary to close some of the loops and study the *partially* closed loop plant. Note that closing the specified loops in the distillation column has made pairing on zero elements in the RGA desirable.

■

Example 3.3.2

Consider the following multivariable plant

$$G(s) = \frac{1-s}{(1+5s)^2} \begin{bmatrix} 1 & -4.19 & -25.96 \\ 6.19 & 1 & -25.96 \\ 1 & 1 & 1 \end{bmatrix}$$

and the corresponding RGA at all frequencies is

$$\Lambda(G) = \begin{bmatrix} 1 & 5 & -5 \\ -5 & 1 & 5 \\ 5 & -5 & 1 \end{bmatrix}$$

In this case, the pairing rules of section 2.3 would recommend a diagonal pairing. It is shown in example 2.3.2 that this gives a very sluggish pairing and $(u_1 - y_3, u_2 - y_1, u_3 - y_2)$ corresponding to 5 in the RGA gives a much faster and better performance.

We now consider two separate cases. In the first case, the diagonal pairing is selected and it is assumed that y_3 is under tight control using u_3. The remaining 2×2 open loop transfer function matrix will be

$$\bar{G}_{12}(s) = \frac{1-s}{(1+5s)^2}\begin{bmatrix} 26.96 & 21.77 \\ 32.15 & 26.96 \end{bmatrix}$$

The corresponding RGA matrix for the partially controlled plant at all frequencies is

$$\Lambda(\bar{G}_{12}) = \begin{bmatrix} 26.98 & -25.98 \\ -25.98 & 26.98 \end{bmatrix}$$

Hence, the large RGA elements imply poor closed loop performance. In the second case, y_3 is paired with u_1 and assuming perfect control in this case the remaining 2×2 open loop transfer function matrix will be

$$\bar{G}_{23}(s) = \frac{1-s}{(1+5s)^2}\begin{bmatrix} -5.19 & -26.96 \\ -5.19 & -32.15 \end{bmatrix}$$

The corresponding RGA matrix for the partially controlled plant at all frequencies is

$$\Lambda(\bar{G}_{23}) = \begin{bmatrix} 6.19 & -5.19 \\ -5.19 & 6.19 \end{bmatrix}$$

which imply the pairing $(u_2 - y_1, u_3 - y_2)$. It is important to note that in this case positive RGA elements are reasonably small, and the RGA for the partially controlled plant does not imply any control difficulty.

■

The above examples show that the RGA may in some cases lead to incorrect pairing selections. In fact, the effect of closing one or more control loops on the properties of the remaining part of the plant is not transparent in the RGA analysis, while, closing the selected loops in a control system can significantly affect the performance of the overall closed loop plant.

Based on the above observations, partially controlled plants can improve the designer's knowledge of the plant under control. By investigating the relative gains of the uncontrolled part of the plant, as in examples 3.3.1 and 3.3.2, better pairing selections can be made. The relative gains of the uncontrolled part are called the *partial relative gains* and are now formally defined.

Definition 3.3. Let some of the loops of the plant $G(s)$ be under integral feedback control. Denote the remaining uncontrolled transfer function matrix by $\bar{G}_R(s)$. Then, the steady state partial relative gain (PRG) for the remaining part is

$$\Lambda_R^P(G) = \Lambda(\bar{G}_R) = \bar{G}_R .* \bar{G}_R^{-T} \tag{3.15}$$

The following theorem gives the necessary conditions for IC (see definition 2.2) with integrity (see section 2.2.3) using the PRG definition. For a proof of the theorem refer to (Haggblom 1997b).

Theorem 3.1

The multivariable plant $G(s)$ is IC with integrity (ICI) only if $\lambda_{ii}(G) > 0$ for $i = 1, \cdots, m$ and $\lambda_{ii}(\bar{G}_{Rk}) > 0,\ i = 1, \cdots, k$ for all principal subsystems, $\bar{G}_{Rk}$ of size $k \times k$, $k = 2, \cdots, m-1$. If in addition to $\lambda_{ii}(G) > 0$, $NI(G) > 0$, the condition is redundant for $k = 2$.

■

Example 3.3.3

Consider the following steady state model of a Petlyuk distillation column (Haggblom 1997b)

$$G = \begin{bmatrix} 153.45 & -179.34 & 0.23 & 0.03 \\ -157.67 & 184.75 & -0.10 & 21.63 \\ 24.63 & -28.97 & -0.23 & -0.10 \\ -4.80 & 6.09 & 0.13 & -2.41 \end{bmatrix}$$

Then

$$\Lambda = \begin{bmatrix} 24.5230 & -23.6378 & 0.1136 & 0.0012 \\ -48.9968 & 49.0778 & 0.0200 & 0.8990 \\ 38.5591 & -38.6327 & 1.0736 & 5\times10^{-6} \\ -13.0852 & 14.1927 & -0.2072 & 0.0998 \end{bmatrix}$$

Let us denote the various possible pairings by C_{wxyz}, where $wxyz$ are digits from 1 to 4 and a digit j in position i in the subscript of C indicates that y_i is paired with u_j. The possible pairings for the plant with positive relative gains and positive NI (see equation (2.41)) are C_{1234}, C_{1342} (λ_{34} is small positive), C_{1432}, C_{3214}, C_{3412}, and C_{4312}, e.g. C_{4312} denotes a pairing $(u_4 - y_1, u_3 - y_2, u_1 - y_3, u_2 - y_4)$. According to the pairing rules of section 2.3, control configurations in the order of preference would be: C_{1432}, C_{3412}, C_{1234}, C_{3214}, C_{1342} and C_{4312}. Note that all of these pairings contain relative gains greater than 24, and thus high performance decentralized control is improbable with any of these configurations. The RGA does not provide any further information to the designer. However, further screening is possible by employing theorem 3.1 and the PRG.

The configurations C_{1234}, C_{1342} and C_{1432} have the pairing $(u_1 - y_1)$. Closing this loop by integral feedback gives the remaining open loop transfer function matrix and the corresponding PRG as

$$\bar{G} = \begin{matrix} \\ y_2 \\ y_3 \\ y_4 \end{matrix} \begin{matrix} u_2 & u_3 & u_4 \\ \left[\begin{matrix} 0.4780 \\ -0.1844 \\ 0.4801 \end{matrix}\right. & \begin{matrix} 0.1363 \\ -0.2669 \\ 0.1372 \end{matrix} & \left.\begin{matrix} 21.6608 \\ -0.1048 \\ -2.4091 \end{matrix}\right] \end{matrix}, \quad \Lambda(\bar{G}) = \begin{bmatrix} 0.1270 & \boxed{-0.0272} & 0.9003 \\ -0.2460 & 1.2460 & 0.0000 \\ 1.1190 & -0.2187 & 0.0997 \end{bmatrix}$$

Theorem 3.1 rules out the pairing C_{1342}, since $(u_3 - y_2)$ has a corresponding negative element in the PRG. Also, note that the pairing C_{1432} is ideal as the PRG has corresponding elements close to unity.

The configurations C_{3214}, C_{3412}, and C_{4312}, have the pairing $(u_1 - y_3)$. Closing this loop by integral feedback gives the remaining open loop transfer function matrix and the corresponding PRG as

$$\bar{G} = \begin{matrix} \\ y_1 \\ y_2 \\ y_4 \end{matrix} \begin{matrix} u_2 & u_3 & u_4 \\ \left[\begin{matrix} 1.1491 \\ -0.7027 \\ 0.4442 \end{matrix}\right. & \begin{matrix} 1.6629 \\ -1.5724 \\ 0.0852 \end{matrix} & \left.\begin{matrix} 0.9230 \\ 20.9898 \\ -2.4295 \end{matrix}\right] \end{matrix}, \quad \Lambda(\bar{G}) = \begin{bmatrix} 0.1498 & 0.8124 & 0.0378 \\ \boxed{-0.1856} & 0.3229 & 0.8627 \\ 1.0359 & -0.1353 & 0.0995 \end{bmatrix}$$

Theorem 3.1 rules out the pairing C_{3214}, since $(u_2 - y_2)$ has a corresponding negative element in the PRG. Also, note that the pairing C_{3412} has corresponding PRG elements close to unity.

The configurations C_{1432}, C_{3412}, and C_{4312} have the pairing $(u_2 - y_4)$. Closing this loop by integral feedback gives the remaining open loop transfer function matrix and the corresponding PRG as

$$\bar{G} = \begin{matrix} & u_1 & u_3 & u_4 \\ y_1 & -12.0983 & 4.0583 & -70.9403 \\ y_2 & -12.0542 & -4.0438 & 94.7412 \\ y_3 & 1.7965 & 0.3884 & -11.5643 \end{matrix}, \quad \Lambda(\bar{G}) = \begin{bmatrix} 1.9334 & 2.0048 & \boxed{-2.9382} \\ -3.7459 & 0.8083 & 3.9376 \\ 2.8125 & \boxed{-1.8131} & 0.0006 \end{bmatrix}$$

Theorem 3.1 rules out the pairings C_{4312}, and C_{1432}.

Obviously, the PRG shows that most of the input-output pairs proposed by the conventional RGA are not ICI. The remaining control configurations are therefore C_{3412} and C_{1234} that satisfy all the PRG tests for ICI.

■

3.4 The Relative Interaction Array

Since, RGA is defined based on individual control loops, it can *only* measure the interaction in individual loops. The interaction measure provided by the RGA is the closeness of the RGA elements to one. Hence, several pairs may satisfy the RGA guideline for input-output pairing and the designer runs into ambiguity in the pairing process. To resolve this issue, some authors have proposed the *overall interaction measure* concept, which claims to lead to the best input-output pairing selection. An algebraic overall interaction measure is proposed in (Mijares et al. 1986). This measure is unable to address the closed loop stability, robustness or integrity issues, and is therefore of less practical importance. In (Zhu and Jutan 1993), an *intuitive* measure of the overall interaction measure is given as follows

$$\min \sum \left| \lambda_{ij}^k - 1 \right| \qquad \forall k \tag{3.16}$$

where λ_{ij}^k denotes the paired RGA elements corresponding to the kth alternative.

Hence, in the case of several pairing alternatives proposed by the RGA, the overall measure given by (3.16), may aid the designer by identifying the pairing with minimal overall interaction. However, as the distance of the RGA elements from one does not quantify the amount of interaction, this measure would suffer from similar limitations to the RGA.

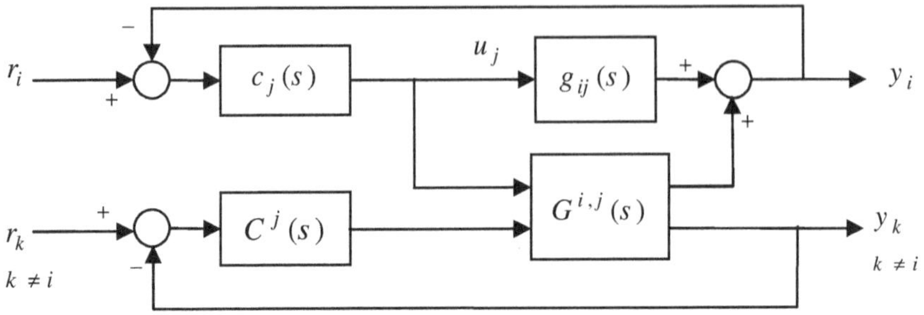

Fig. 3.7 The closed loop plant (The $G^{i,j}(s)$ and $C^j(s)$ are obtained by removing the $(i,j)^{th}$ element of $G(s)$ and j^{th} row of $C(s)$ respectively).

Interaction analysis. Consider the $m \times m$ transfer function matrix $G(s)=[g_{ij}(s)]$, the corresponding $m \times m$ compensator matrix $C(s)=diag\{c_j(s)\}$ and the pair $(u_j - y_i)$. To study the plant interactions, the block diagram of the closed loop plant can be redrawn as shown in figure 3.7. The direct path from u_j to y_i is *interaction-free*, and the parallel path contains all the interactions resulting from the other loops, that is called the *absolute loop interaction*. Thus, we have

$$y_i(s)=[g_{ij}(s)+a_{ij}(s)]u_j(s) \tag{3.17}$$

where $a_{ij}(s)$ can be evaluated by simple signal flow graph technique or direct algebraic manipulations. The *relative interaction* is now defined as the ratio of the absolute interaction and the interaction free transfer function as (Zhu 1996)

$$\phi_{ij}(s)=\frac{a_{ij}(s)}{g_{ij}(s)} \tag{3.18}$$

Hence, equation (3.17) gives

$$y_i(s)=g_{ij}(s)[1+\phi_{ij}(s)]u_j(s)=\tilde{g}_{ij}(s)u_j(s) \tag{3.19}$$

The following points are worth noting:

- The amplification or attenuation factor $1+\phi_{ij}(s)$, measures the deviation from the interaction free transfer function.
- Equation (3.19) can be modeled as a multiplicative uncertainty representation.

- No assumptions are made about the plant or the decentralized control scheme.
- By assuming the closed loop integrity and integral control action in each loop, at steady state we have $a_{ij} = \frac{1}{[G^{-1}]_{ji}} - g_{ij}$.

3.4.1 The Relative Interaction Array and the Pairing Rules

The relative interaction array (RIA) can be defined using equation (3.18) and the steady state values of a_{ij}. The following materials are based on (Zhu 1996).

Definition 3.4. The RIA is defined as $\Phi_{RIA} = \left[\phi_{ij}\right]$, where

$$\phi_{ij} = \frac{1}{\lambda_{ji}} - 1 \tag{3.20}$$

and λ_{ij} is the element of the corresponding RGA. Hence, it is easily seen that

$$\Phi_{RIA} = \left[\phi_{ij} = \frac{1}{\lambda_{ij}} - 1\right] \tag{3.21}$$

and also

$$\lambda_{ij} = \frac{1}{\phi_{ij} + 1} \tag{3.22}$$

Equations (3.19) and (3.22), give

$$\tilde{g}_{ij} = (1 + \phi_{ij}) g_{ij} = \frac{1}{\lambda_{ij}} g_{ij} \tag{3.23}$$

which is the equivalent plant gain. The following observations now follow:

- $\phi_{ij} = 0$ implies no interaction and $\lambda_{ij} = 1$.
- $\phi_{ij} > 0$ implies that interaction is acting in the same direction as the interaction -free plant gain.
- $\phi_{ij} > 1$ implies interaction is dominating the interaction-free plant gain.
- $\phi_{ij} < 0$ implies that interaction is acting in the reverse direction of the interaction-free plant gain.

- $\phi_{ij} < -1$ implies that reverse interaction is dominating the interaction-free plant gain. Also, ϕ_{ij} close to -1 is equivalent to very large positive or negative λ_{ij}, i.e. large RGA elements.

Necessary stability condition requires that $\tilde{g}_{ij}c_j > 0$ and $g_{ij}c_j > 0$ for all j, where c_j is the compensator steady state gain. This condition and equation (3.23) give the stability necessary condition in terms of the elements of the RIA as $\phi_{ij} > -1$ for all j.

Note. The necessary stability condition in terms of the elements of the RIA is equivalent to (see equation (3.22)) $\lambda_{ij} > 0$ for all j, that is the RGA integrity condition. It is also consistent with the NI stability condition.

The RIA based pairing rules. Based upon the above observations the following pairing rules are proposed:

- Select pairings for which the RIA elements are closest to zero.
- Select pairings for which the NI is positive.
- Select pairings for which the RIA elements are greater than -1.
- Avoid pairing that gives RIA elements close to -1.

An overall interaction measure based on the RIA. Similar to the RGA, several pairing candidates may satisfy the above RIA pairing rules. To choose the best candidate, the following overall interaction measure is introduced

$$\min \sum \left| \phi_{ij}^{k} \right| \qquad \forall k \tag{3.24}$$

where ϕ_{ij}^{k} denotes the paired RIA elements corresponding to the kth alternative. In terms of the RGA elements equation (3.24) becomes

$$\min \sum \left| \frac{1}{\lambda_{ij}^{k}} - 1 \right| \qquad \forall k \tag{3.25}$$

Comparing equations (3.16) and (3.25), shows that very small RGA elements corresponding to large interactions are explicitly penalized in equation (3.25), while very large RGA elements are neglected due to robustness considerations. Hence, the following can be added to the above pairing rules

- Select pairings for which, $\sum \left| \phi_{ij}^{k} \right|$ is minimum.

Example 3.4.1

Consider the following steady state gain matrix (Zhu 1996)

$$G(0) = \begin{bmatrix} 1.0 & 1.0 & -0.1 \\ 1.0 & -3.0 & 1.0 \\ 0.1 & 2.0 & -1.0 \end{bmatrix}$$

Then the RGA and RIA matrices are given below

$$\Lambda = \begin{bmatrix} 0.53 & 0.59 & -0.12 \\ 0.43 & 1.59 & -1.02 \\ 0.04 & -1.18 & 2.14 \end{bmatrix}$$

and

$$\Phi_{RIA} = \begin{bmatrix} 0.87 & 0.70 & -9.13 \\ 1.34 & -0.37 & -1.98 \\ 25.71 & -1.85 & -0.53 \end{bmatrix}$$

The pairings proposed by the RGA and the RIA are $(u_1 - y_1, u_2 - y_2, u_3 - y_3)$, and $(u_1 - y_2, u_2 - y_1, u_3 - y_3)$. However, RGA can not go any further in introducing the best choice. The overall interaction measures for the pairings using equation (3.24) are 1.77 and 2.57, respectively. Hence, the RIA pairing rules suggest the first set as the best pair.

■

3.5 Block Pairing and Block Relative Gain

Decentralized control systems are widely used in industrial plants and process industries. The benefits and advantages of decentralized control, and the main reasons behind their extensive applications were discussed in section 1.1. However, in the face of severe interactions, the decentralized control of highly interactive multivariable plants can result in poor closed loop performance. To overcome this interaction problem, centralized controllers are an alternative. But, the costs and difficulties in employing fully centralized control systems, as discussed in section 1.1, has resulted in their limited use. On the other hand, in many industrial plants and process industries interaction is not heavily distributed through the plant, it is sever in some parts and less disturbing in the others. A midway between fully decentralized and centralized control systems is the use of *block decentralized controllers*. Block decentralized controllers inherit the simplicity of implementation and maintenance of decentralized control and high performance of centralized controllers.

3.5.1 A Brief Review of the Methods

In 1986, the concept of RGA was generalized to Block Relative Gain (BRG) in (Manousiouthakis et al. 1986). In contrast to the RGA that gives SISO input-output pairings, BRG provides a block pairing of inputs and outputs structures. In this generalized framework, different types of subsystems with given sets of outputs and inputs are available which yield various block decentralized controllers. However, the number of candidate blocks and hence possible designs increases with the plant dimensions. This necessitates the use of BRG to give an estimate of the closed loop plant behavior prior to the design.

By assuming *perfect* control, the initial BRG failed to provide a complete direct theoretical link between block decompositions, closed loop stability and performance. Although limited, some progress was made to establish this link in (Nett and Manousiouthakis 1987), and (Manousiouthakis and Arkun 1987). The results of (Nett and Manousiouthakis 1987) identifies potential control problems by evaluating the spectral radius of the BRG and show that the spectral radius of the plant BRG is a lower bound for the plant condition number, that is the ratio between maximum and minimum singular values of the transfer function matrix, and (Manousiouthakis and Arkun 1987) establishes connections between BRG and generalized block diagonal dominance. In (Chen 1992), relations between BRG and the condition number are further elaborated and improved. It is shown that the maximum singular value and the structured singular value of the BRG are the lower bounds for the condition number and minimal condition number, respectively (Chen 1992). Hence, large maximum singular value and large structured singular value of the BRG indicates difficulty in the control of the plant.

Dynamic Block Relative Gain (DBRG) is introduced by (Arkun 1987). DBRG is developed using the Decentralized Model Reference Schemes (DMRS). The closed loop performance is studied by defining the performance deterioration index, which comes from the difference between the achievable performance when subsystem interactions are ignored and the actual performance in the presence of interactions. A theorem is also provided to establish the link between DBRG and closed loop stability. These results are dependent upon the block controller designs.

Despite its constructive generalization of BRG, DBRG fails to provide practical indications of the magnitude and direction of the dynamic interaction among the blocks in the multivariable plant. In addition, due to its dependency on the controller designs, it is not easy to get the proper input-output pairings. To overcome these problems, relative sensitivity is introduced in (Arkun 1988) as a dynamic closed loop interaction measure (IM). Relative sensitivities are defined for a particular subsystem and its complements in the plant transfer function matrix in the block partition form. Using the computed relative sensitivities, closed loop dynamic interactions are derived. Subsequently, pairing and design rules are provided.

Nonsquare Block Relative Gain (NBRG) developed in (Reeves and Arkun 1989) generalizes the BRG concept to nonsquare multivariable plants. It is assumed that if the plant transfer function matrix has more outputs than inputs, then its blocks must either be square or have more outputs than inputs. Conversely, if it has more inputs than outputs, its diagonal blocks must have the number of inputs greater than or equal to the number of outputs. Also, the plant transfer function matrix and its diagonal blocks must have full rank. Perfect control is assumed through the derivation of NBRG and least squares or minimum-norm solutions will be used in cases where exact solutions are not possible. (Reeves and Arkun 1989) also gives the definition of Nonsquare DBRG (NDBRG). And the closed loop performance, stability, and relative sensitivity issues are extended to the nonsquare case (For NSRGA see section 2.4.1).

In a recent comprehensive paper (Kariwala et al. 2003), BRG concept is studied and its various properties are derived. Some algebraic properties of the BRG, and the important issues such as, stability, controllability, block diagonal dominance and interactions are dealt with. The closed loop properties are derived with the assumption of integral control in all loops. Hence, integral controllable with integrity (ICI) and block decentralized integral controllable (block-DIC) are defined. Due to the importance of the ICI and block-DIC issues, they are the subject of many independent papers. A good treatment using the NI and BRG notions can be found in (Kariwala et al. 2005). A proposition is provided in (Kariwala et al. 2003) that identifies non-minimum phase transmission zeros and some properties are derived. The claim in (Manousiouthakis et al. 1986) that BRG closer to the identity matrix means weaker interactions is falsified in (Kariwala et al. 2003). It is shown that a plant with BRG equal the identity matrix can indeed have a high degree of interactions. But, singular values of the BRG are employed to study the interaction problem and it is shown that closed loop large interactions exists if the singular values are far from unity or alternatively, BRG is far from the identity matrix, but the converse is not true. Finally, four pairing rules are given that facilitate the input-output *block* pairing problem (Kariwala et al. 2003).

3.5.2 Fundamental Results in Block Relative Gain and Block Pairing

The BRG related definitions and some key results are provided in this section based on the results of (Kariwala et al. 2003). Let $G(s)=[g_{ij}(s)]$ be an $m \times m$ transfer function matrix. Then, for block decentralized control, it is desired to decompose the original plant into M non-overlapping square sub plants such that,

$G_{ii}(s)$ are $m_i \times m_i$ transfer function matrices and $i = 1, 2, \cdots, M$; $\sum_{i=1}^{M} m_i = m$. Also, $G_{ij}(s)$ an $m_i \times m_j$ transfer function matrix, is the block gain relating the vector input and output pair $(\mathbf{u}_j - \mathbf{y}_i)$, where in this section boldface letter is used to emphasis the vector of output and input variables in block pairings.

Definition 3.5. The BRG for the pair $(\mathbf{u}_1 - \mathbf{y}_1)$ is defined as the ratio of the open loop block gain and apparent block gain in the same loop when all other control loops are closed

$$[\Lambda_B(s)]_{11} = G_{11}(s)[G^{-1}(s)]_{11} \tag{3.26}$$

where $G_{11}(s)$ is the $m_1 \times m_1$ transfer function matrix and $m_1 \leq m$, relating the first m_1 inputs and outputs of $G(s)$, and $[G^{-1}(s)]_{11}$ is the corresponding block of $G^{-1}(s)$.

Note. Definition 3.5 defines the *left* BRG. The *right* BRG is defined similarly. Since, right and left BRG have similar properties, only the left BRG is considered and it is simply called the BRG.

Let the multivariable plant described by the transfer function matrix $G(s)$ be partitioned in the following form to conform with definition 3.4

$$\begin{aligned} \mathbf{y}_1(s) &= G_{11}(s)\mathbf{u}_1(s) + G_{12}(s)\mathbf{u}_2(s) \\ \mathbf{y}_2(s) &= G_{21}(s)\mathbf{u}_1(s) + G_{22}(s)\mathbf{u}_2(s) \end{aligned} \tag{3.27}$$

assuming perfect control for $(\mathbf{u}_2 - \mathbf{y}_2)$, we have at steady state

$$\mathbf{y}_1 = \bar{G}_{11}\mathbf{u}_1; \qquad \bar{G}_{11} = G_{11} - G_{12}G_{22}^{-1}G_{21} \tag{3.28}$$

For partitioned matrices, $\left[G^{-1}\right]_{11} = \bar{G}_{11}^{-1}$. Hence, the steady state BRG between $(\mathbf{u}_1 - \mathbf{y}_1)$ is

$$[\Lambda_B]_{11} = G_{11}\bar{G}_{11}^{-1} = G_{11}[G^{-1}]_{11} \tag{3.29}$$

For the general case that G is partitioned into M diagonal and conformal off diagonal blocks, such that $G_{ii} \in \Re^{m_i \times m_i}$ $i = 1, 2, \cdots, M$, we have

$$[\Lambda_B]_{ii} = G_{ii}\bar{G}_{ii}^{-1} = G_{ii}[G^{-1}]_{ii} \tag{3.30}$$

The BRG properties. Similar to the RGA, certain algebraic properties can be proved for the BRG. These properties are summarized below. The detailed proof can be found in (Kariwala et al. 2003):

- The BRG elements are a weighted sum of the RGA elements as $\beta_{ij} = \sum_{k=1}^{m_1} \frac{g_{ik}}{g_{jk}} \lambda_{jk}$, where $\left[\Lambda_B\right]_{11} = \left[\beta_{ij}\right] \in \Re^{m_1 \times m_1}$.

The first RGA property states that the row or column sum of the RGA elements add up to one. A similar property can be derived for the BRG:

- Let P be a m_1-dimensional ordered index set chosen form the first m natural numbers, $\tilde{G}$ be obtained from permutation of G such that $\tilde{G}_{11} = G_{pq}$, where G_{pq} is a submatrix of G made up of its rows and columns indexed by p and q, respectively. Then $\mathbf{y}_p \subset \mathbf{y}$, $\mathbf{u}_q \subset \mathbf{u}$ and $[\Lambda_B(\tilde{G})]_{11}$ is the BRG between $\mathbf{y}_p$ and $\mathbf{u}_q$. Then for any $p \subset P$

$$\sum_{q \subset P} [\Lambda_B(\tilde{G})]_{11} = \frac{m_1}{n} \frac{n!}{m_1!(n-m_1)!} I_{m_1} \tag{3.31}$$

- Let the plant gain matrix be scaled as $G^s = S_1 G S_2$, where S_1 and S_2 are real diagonal output and input scaling matrices respectively. If the scaling matrices are portioned as $S_1 = diag(S_{11}, S_{12})$ and $S_2 = diag(S_{21}, S_{22})$, where S_{11}, $S_{21} \in \Re^{m_1 \times m_1}$, it can be shown that

$$\left[\Lambda_B^s\right]_{11} = S_{11}\left[\Lambda_B\right]_{11} S_{11}^{-1} \tag{3.32}$$

Equation (3.32) shows that the BRG is input scaling independent and output scaling dependent (compare with the third RGA property in section 2.2.2), and $[\Lambda_B]_{ii}$ is independent of scaling of $[\Lambda_B]_{jj}$ $\forall i, j = 1, \cdots, M$, $i \neq j$. Also, if all the outputs are scaled by the same scalar, that is $S_{11} = sI_{m_1}$ the BRG would be output scaling independent. Finally, the diagonal elements of the BRG are independent of scaling.

- The respective BRG exists and is singular for nonsingular G and *singular* G_{ii}.
- For some specific partitioning, block triangular plants can imply corresponding identity matrix BRG, but the converse is not true.

Some closed loop properties. Consider the closed loop multivariable plant shown in figure 3.8. It is assumed that $K_{ii}(s)=\left(\frac{k_i}{s}I_{m_i}\right)C_{ii}(s)$, $k_i>0$ and $P(s)=G(s)C(s)$ is stable and proper. Two desired closed loop properties are now stated in terms of the following definitions, which are the block versions of the similar definitions.

Definition 3.6. The plant $G(s)$ is called *integral controllable with integrity* (ICI) if the block decentralized controllers of figure 3.8 exits to ensure closed loop stability when the gain of all individual controllers are simultaneously detuned by a factor ε, $0<\varepsilon\leq 1$, and also any combination of the individual controllers are brought in and out of service.

Definition 3.7. The plant $G(s)$ is called *block decentralized integral controllable* (block-DIC) if the block controller $K(s)$ of figure 3.8 exits to ensure closed loop stability when the gain of the respective block controllers are independently detuned by $0\leq\varepsilon_i\leq 1$, $i=1,2,\cdots,M$.

The following stability results are now stated, to establish the link between the BRG and the closed loop properties.

Lemma 3.1

Consider the closed loop plant shown in figure 3.8, with $k_1=k_2=\cdots=k_M$. Then, the closed loop plant is stable *only if* $\det(P(0))>0$.

■

For the proof see (Grosdidier and Morari 1986).

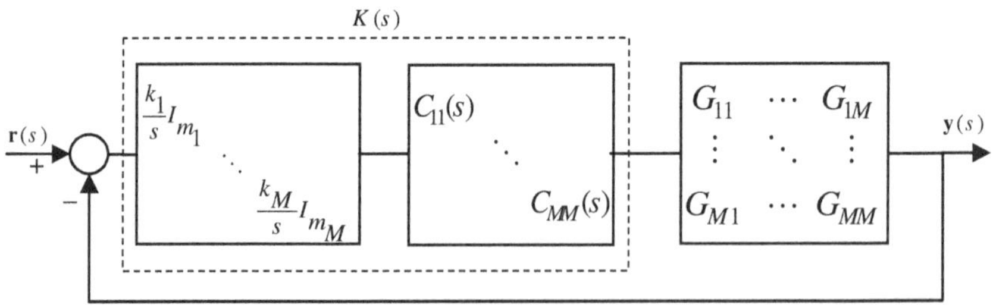

Fig. 3.8 Block diagram of the closed loop multivariable plant.

Theorem 3.2

Consider the closed loop plant shown in figure 3.8. If $\det\left(\left[\Lambda_B(0)\right]_{ii}\right) < 0$ for some i, $i = 1, 2, \ldots, M$, then at least one of the following conditions hold:

- The closed loop plant is unstable.
- The closed loop plant with the ith loop removed is unstable.
- The ith loop is unstable on its own.

■

Theorem 3.3

Consider the closed loop plant shown in figure 3.8. Let all the individual loops be stable with zero steady state error. Further assume that $\det(G_{ii}(0)) \neq 0$ for all $i = 1, \ldots, M$. Then, the closed loop plant is stable *only if*

$$\text{GNI} = \frac{\det(G(0))}{\prod_{i=1}^{M} \det(G_{ii}(0))} > 0 \tag{3.33}$$

■

For the proofs of the above theorems see (Kariwala et al. 2003).

Note. Equation (3.33) defines the Generalized Niederlinski Index (GNI).

Non-minimum phase behaviors. The property 10 in section 2.2.2 for the RGA is extended to the BRG in (Kariwala et al. 2003). These results are summarized without proof. The proofs are given in (Kariwala et al. 2003).

Result 1. Consider a stable multivariable plant decomposed as given by equations (3.27). If there exits m_1, $2 \leq m_1 \leq m-2$ such that $\det\left(\left[\Lambda_B(j\infty)\right]_{11}\right)$ is nonzero, finite and has a different sign from $\det\left(\left[\Lambda_B(0)\right]_{11}\right)$, then either $G_{11}(s)$ or $G_{22}(s)$ will have a RHP-transmission zero.

Result 2. If $G_{22}(s)$ contains a RHP-transmission zero, and all the loops except $(\mathbf{u}_1 - \mathbf{y}_1)$ are closed, then the open loop subsystem $(\mathbf{u}_1 - \mathbf{y}_1)$ or $\bar{G}_{11}(s)$ contains a RHP-pole.

Result 3. If there exits m_1, $1 \le m_1 \le m-1$ such that $\det\left(\left[\Lambda_B(0)\right]_{11}\right)=0$, then one or both of $G_{11}(s)$ and $G_{22}(s)$ will have a transmission zero at the origin.

Before proceeding to the next topic, an important assertion proved in (Kariwala et al. 2003) is that *if the BRG is far from the identity matrix the plant has large interactions.* However, the converse is not true.

Block diagonal dominance. Block diagonal or triangular multivariable plants are easy to control, since each controller can be tuned independently. But, this is not the case in most practical plants. However, if the effect of a control signal or a set of control signals is larger than the effect of the other control signals on a particular output or set of outputs, then independent tuning is feasible. The way to evaluate these interaction effects and their size is to use the *block diagonal dominance* concept.

Definition 3.8 The matrix Z is called *generalized row block diagonal dominant* if there exists $x \in \Re^M$, $x > 0$ such that for a given partitioning,

$$\left\|Z_{ii}^{-1}\right\|^{-1} x_i > \sum_{j=1, j\neq i}^{M} \left\|Z_{ij}\right\| x_j \qquad i=1,2,\ldots,M \tag{3.34}$$

If x is chosen as $\begin{bmatrix}1 & 1 & \cdots & 1\end{bmatrix}^T$, then Z is called row block diagonal dominant. If Z is generalized block diagonal dominant (GBDD), there exists a scaling matrix $X = diag(x_i . I_{m_i})$ such that XZX^{-1} is block diagonal dominant.

Many interesting results can be derived using the GBDD definition. We will review a series of results relevant to the block pairing problem, all proved in (Kariwala et al. 2003).

Result 4. Suppose that a compensator matrix $C(s)$ is designed for $G(s)$ such that $P(s)=G(s)C(s)$ is stable and proper. It can be shown that, *if* $P(0)$ *is GBDD then* $G(s)$ *is block-DIC.*

This is an attractive result but its application is limited since it is compensator dependent. A compensator independent result which uses the BRG concept is given next.

Result 5. $P(s)$ is GBDD *only if* $\bar{\sigma}\left(\left[\Lambda_B(0)\right]_{ii}\right) > 0.5$ for all $i=1,2,\cdots,M$.

A key property of the GBDD plants is the fact that the stability of individual loops implies the stability of the closed loop plant. However, application of GBDD leads to conservative results and one way to partially solve the conservativeness is the use of *quasi block diagonal dominance* (QBDD). The key stability property of GBDD also holds for QBDD. This makes it attractive for independent loop tuning and decentralized control.

Definition 3.9. The matrix Z is called quasi block diagonal dominant if there exists $x \in \Re^M$, $x > 0$ such that for a given partitioning

$$x_i > \sum_{j=1, j \neq i}^{M} \left\| Z_{ij} Z_{ii}^{-1} \right\| x_j \qquad i = 1, 2, \ldots, M; Z_{ii} \neq 0 \tag{3.35}$$

Result 6. $P(s)$ is *QBDD only if* $\bar{\sigma}\left(\left[\Lambda_B(0)\right]_{ii}\right) > 0.5$ for all $i = 1, 2, \cdots, M$.

Consider the block diagram shown in figure 3.9. Let $P_{bd}(s)$ and $G_{bd}(s)$ be the matrices containing the block diagonal elements of $P(s)$ and $G(s)$ respectively and define the matrix $E(s) = \left(P(s)P_{bd}^{-1}(s) - I_m\right) = \left(G(s)G_{bd}^{-1}(s) - I_m\right)$. Then the following important results hold.

Result 7. $P(s)$ is GBDD *only if* $\mu_\Delta(E(0)) < 1$. Where, $\mu_\Delta(E(s))$ is the structured singular value of $E(s)$.

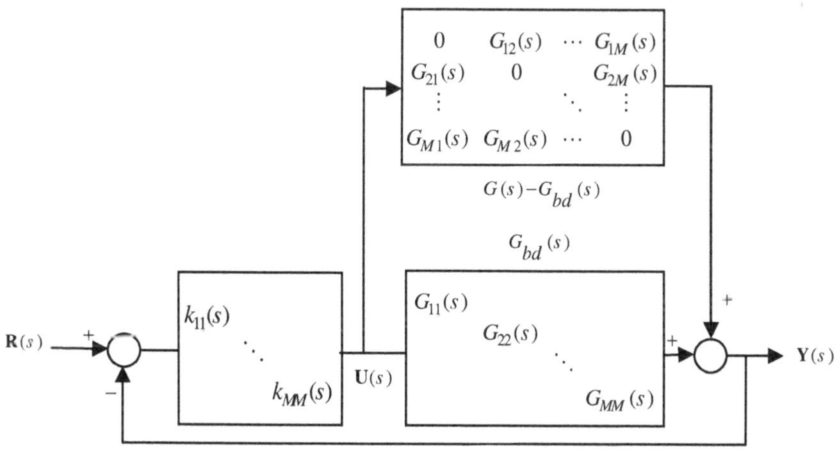

Fig. 3.9 Plant decomposition into block diagonal and off-block diagonal elements.

Result 8. For a proper $G(s)$, $\mu_\Delta(E(0))<1$ *only if* $\bar{\sigma}\left([\Lambda_B(0)]_{ii}\right)>0.5$ for all $i=1,2,\cdots,M$.

Interaction analysis. Consider the block diagram shown in figure 3.9. The ideal block decoupled non interacting case is $G(s)=G_{bd}(s)$. Define $\Gamma(s)=G_{bd}(s)G^{-1}(s)$ as the *performance relative gain array* (PRGA). We can argue that $\sigma_i(\Gamma(s))\approx 1$ for all i, leads to low interaction in the multivariable plant. This is equivalent to the minimization of $\sum_i \left|\sigma_i(\Gamma(0))-1\right|$.

Note. The idea of Performance Relative Gain Array (PRGA) and its properties are given in (Hovd and Skogestad 1992). PRGA is defined as follows

$$\Gamma(s)=\hat{G}(s)G(s)^{-1} \tag{3.36}$$

where $\hat{G}(s)$ is the matrix consisting of only the diagonal elements of $G(s)$. This new definition is developed to overcome the deficiency of the original RGA in coupling identification in certain multivariable plants, e.g. the triangular multivariable plants. Although the diagonal elements of the RGA and PRGA are the same, their algebraic properties are different. PRGA is dependent on row and column permutations and output scaling, but independent of input scaling.

Block pairing rules. The above results lead to the following pairing rules:

- Avoid pairing that gives $\det([\Lambda_B(s)]_{ii})\le 0$ for some $i=1,\cdots,M$ or $NI<0$.
- Avoid pairing that give different signs of $\det([\Lambda_B(0)]_{ii})$ and $\det([\Lambda_B(j\infty)]_{ii})$.
- Prefer pairings for which, $\mu_\Delta(E(0))<1$.
- Prefer pairings for which, $\sum_i \left|\sigma_i(\Gamma(0))-1\right|$ is small.

Note. Considering fully decentralized controllers as the simplest multivariable control structure, $\sum_i \left|\sigma_i(\Gamma(0))-1\right|$ tends to be smaller for structures approaching the fully centralized structure. Hence, the designer is confronted with a compromise between the controller structure complexity and closed loop performance.

Example 3.5.1

Consider the Alstom gasifier plant with scaled dc gain matrix as follows (Kariwala et al. 2003)

$$G(0)=\begin{bmatrix} 0.0385 & -0.0427 & 0.0444 & -0.0474 \\ -0.1115 & -0.0297 & 0.0770 & -0.0142 \\ 0.0327 & 0.8630 & 0.0477 & 0.5019 \\ 0.0088 & 0.1284 & -0.1101 & -0.2834 \end{bmatrix}$$

For block decentralize control we assume $m_1=3$ and $m_2=1$. Studying the 16 possible different cases, reveals that only 2 cases satisfy all the required conditions. These pairings are shown in Table 3.1. Therefore, $(u_{2,3,4}-y_{1,3,4},u_1-y_2)$ is the appropriate block pair with lower interaction due to the minimum value of $\sum_i \left|\sigma_i(\Gamma(0))-1\right|$.

■

Table 3.1 Block pairing analysis for Alstom gasifier plant.

	Block Pairing	$\sum_i \left\|\sigma_i(\Gamma(0))-1\right\|$
1	$(u_{1,3,4}-y_{1,2,4},u_2-y_3)$	1.8608
2	$(u_{2,3,4}-y_{1,3,4},u_1-y_2)$	0.8965

Example 3.5.2

The column/stripper distillation plant transfer function matrix is as follows (Kariwala et al. 2003)

$$G(s)=\begin{bmatrix} \dfrac{4.09e^{-1.3s}}{(33s+1)(8.3s+1)} & \dfrac{-6.36e^{-1.2s}}{(31.6s+1)(20s+1)} & \dfrac{-0.25e^{-1.4s}}{(21s+1)} & \dfrac{-0.49e^{-6s}}{(22s+1)^2} \\ \dfrac{-4.17e^{-5s}}{(45s+1)} & \dfrac{6.93e^{-1.02s}}{(44.6s+1)} & \dfrac{-0.05e^{-6s}}{(34.5s+1)^2} & \dfrac{1.53e^{-3.8s}}{(48s+1)} \\ \dfrac{1.73e^{-18s}}{(13s+1)^2} & \dfrac{5.11e^{-12s}}{(13.3s+1)^2} & \dfrac{4.61e^{-1.01s}}{(18.5s+1)} & \dfrac{-5.49e^{-1.5s}}{(15s+1)} \\ \dfrac{-11.2e^{-2.6s}}{(43s+1)(6.5s+1)} & \dfrac{14(10s+1)e^{-0.02s}}{(45s+1)(17.4s^2+3s+1)} & \dfrac{0.1e^{-0.05s}}{(31.6s+1)(5s+1)} & \dfrac{4.49e^{-0.6s}}{(48s+1)(6.3s+1)} \end{bmatrix}$$

It can be easily shown that no block pairing satisfy the pairing rules for the 2×2 blocks. Hence, we assume $m_1=3$ and $m_2=1$. The block pairings satisfying the above rules are shown in table 3.2. Also, the $(u_{1,2,4}-y_{1,2,4},u_3-y_3)$ block pair has the minimum interaction. We further investigate the frequency responses of

$\sum_i \left|\sigma_i\left(\Gamma(s)\right)-1\right|$ for the candidate block pairs in table 3.2. These frequency responses are shown in figure 3.10. It is observed that at low frequencies $(u_{1,2,4}-y_{1,2,4}, u_3-y_3)$ is the appropriate block pair but at high frequencies $(u_{1,3,4}-y_{1,3,4}, u_2-y_2)$ is the appropriate block pair.

■

Table 3.2 Block pairing analysis for column/stripper plant.

	Block Pairing	$\sum_i \left\|\sigma_i\left(\Gamma(0)\right)-1\right\|$
1	$(u_{1,2,4}-y_{1,2,4}, u_3-y_3)$	5.6485
2	$(u_{1,3,4}-y_{1,3,4}, u_2-y_2)$	11.5336

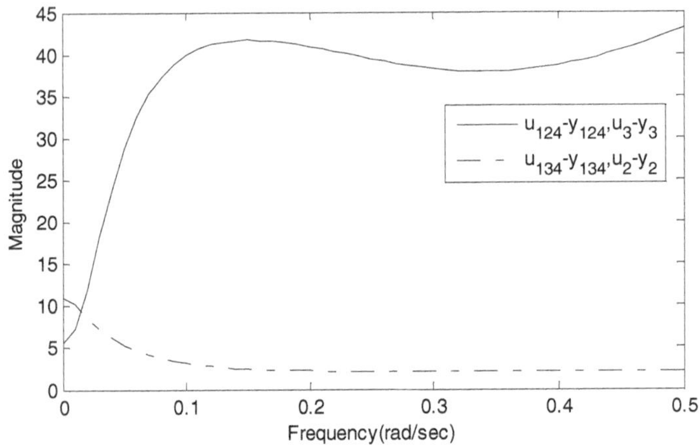

Fig. 3.10 Frequency responses of $\sum_i \left|\sigma_i\left(\Gamma(s)\right)-1\right|$ for the 2 appropriate block pairs.

3.6 Conclusion

The main purpose of this chapter has been to provide an overview of the advanced pairing techniques which are mainly generalizations of the RGA. These include the Dynamic Relative Gain Array, the Effective Relative Gain Array, the Partial Relative Gain, the Relative Interaction Array, Performance Relative Gain Array, Block Pairing and the Block Relative Gain. Numerous examples from the literature are used to show the main points of the pairing methodologies.

References

Arkun, Y.: Dynamic block relative gain and its connection with the performance and stability of decentralized control structure. Int. J. Control 46, 1187–1193 (1987)

Arkun, Y.: Relative sensitivity: a dynamic closed-loop interaction measure and design tool. AIChE J. 34, 672–675 (1988)

Bristol, E.: Recent results on interactions in multivariable process control. In: Proceeding of the 71st Annual AIChE meeting, Houston, TX, USA (1979)

Chen, J.: Relations between block relative gain and Euclidean condition number. IEEE Trans. Autom. Control 37, 127–129 (1992)

Gagnepain, J.P., Seborg, D.E.: Analysis of process interactions with application to multiloop control system design. Ind. Eng. Chem. Proc. Des. Dev. 21, 5–11 (1982)

Grosdidier, P., Morari, M.: Interaction measures for systems under decentralized control. Automatica 22, 309–319 (1986)

Haggblom, K.E.: Control Structure Analysis by Partial Relative Gains. In: Proceeding of the 36th Conference on Decision and Control, San Diego, California, USA (1997a)

Haggblom, K.E.: Partial relative gain: a new tool for control structure selection. In: AIChE Annual Meeting, November 16-21 (1997b)

Hovd, M., Skogestad, S.: Simple frequency-dependent tools for control system analysis, structure selection and design. Automatica 28, 989–996 (1992)

Huang, H.P., Ohshima, M., Hashimoto, I.: Dynamic interaction and multiloop control system design. J. Process Contr. 4, 15–27 (1994)

Kariwala, V., Forbes, J.F., Meadows, E.S.: Block relative gain: properties, and pairing rules. Ind. Eng. Chem. Res. 42, 4564–4574 (2003)

Kariwala, V., Forbes, J.F., Meadows, E.S.: Integrity of systems under decentralized integral control. Automatica 41, 1575–1581 (2005)

Kariwala, V., Forbes, J.F., Skogestad, S.: [graphics object to be inserted manually]-Interaction measure for unstable systems. Int. J. Automation and Control 1, 295–313 (2007)

Levine, W., Athans, M.: On the determination of the optimal constant feedback gains for linear multivariable systems. IEEE Trans. Autom. Control 15, 44–48 (1970)

Lewis, F.L.: Applied Optimal Control & Estimation. Prentice Hall, New Jersey (1992)

Manousiouthakis, V., Arkun, Y.: A Hybrid Approach for the Design of Robust Control Systems. Int. J. Control 45, 2203–2220 (1987)

McAvoy, T.J.: Some results on dynamic interaction analysis of complex control systems. Ind. Eng. Chem. Process Des. Dev. 22, 42–49 (1983)

McAvoy, T., Arkun, Y., Chen, R., Robinson, D., Schnelle, P.D.: A new approach to defining a dynamic relative gain. Control Eng. Pract. 11, 907–914 (2003)

Mijares, G., Cole, C.D., Naugle, N.W., Preisig, H.A., Holland, C.D.: A new criterion for the pairing of control and manipulated variables. AIChE J. 32, 1439–1449 (1986)

Morari, M.: Operability Measures for Process Design. Ind. Chem. Eng. Symposium Series, 131–140 (1982)

Nett, C.N., Manousiouthakis, V.: Euclidean condition and block relative gain: connections, conjectures, and clarifications. IEEE Trans. Autom. Control 32, 405–407 (1987)

Reeves, D.E., Arkun, Y.: Interaction measures for non-square decentralized control structures. AIChE J. 35, 603–613 (1989)

Shinskey, F.: Process Control Systems, ch. 7. McGraw-Hill, New York (1967)

Tung, L.S., Edgar, T.F.: Analysis of control-output interactions in dynamic systems. AIChE J. 27, 690–693 (1981)

Witcher, M.F., McAvoy, T.J.: Interacting control systems: steady-state and dynamic measurement of interaction. ISA T 16, 35–41 (1977)

Xiong, Q., Cai, W.-J.: Effective transfer function method for decentralized control system design on multi-input multi-output processes. J. Process Contr. 16, 773–784 (2006)

Xiong, Q., Cai, W.-J., He, M.J.: A practical loop pairing criterion for multivariable processes. J. Process Contr. 15, 741–747 (2005)

Zhu, Z.X., Jutan, A.: A new variable pairing criterion based on Niederlinski index. Chem. Eng. Commun. 121, 235–250 (1993)

Zhu, Z.X.: Variable pairing selection based on individual and overall interaction measures. Ind. Eng. Chem. Res. 35, 4091–4099 (1996)

Chapter 4
Control Configuration Selection of Linear Multivariable Plants: SSV and Passivity Based Techniques

4.1 Introduction

The advanced RGA based techniques introduced in chapter 3 provide powerful extensions to the well established RGA methodologies. Each of the proposed RGA extensions solves a problem not considered by the static classical RGA. This chapter considers two advanced approaches for control configuration selection that are not RGA based. These techniques are:

- **SSV Approach to Input-Output Pairing.** In this approach, Structured Singular Values (SSV) is employed to quantify the interaction in linear multivariable plants with diagonal or block diagonal controllers. Closed loop stability and performance under decentralized control corresponding to a selected control structure is also considered. The main ideas are given in section 4.2.
- **Passivity based Approach to Control Configuration Selection.** This recently developed approach is presented in section 4.3. This strategy employs the passivity concept. Although the properties and applications of passivity theory are well known in stability analysis and control systems design, it has been newly introduced in interaction analysis and it uses the so-called *degree of passivity*.

4.2 SSV Approach to Input-Output Pairing

Some authors have used a Singular Value (SV) framework to develop new interaction measures (IM) with sound theoretical basis. SVs or the principal gains of the open loop transfer function matrix were initially used to evaluate the multivariable stability margins, and later adopted for loop selection in a steady state plant (Smith et al. 1981), (Smith 1981) and possible achievable performance with sensitivity analysis in multivariable plants (Morari 1982). In (Lau et al. 1985), SV methodology is extended to the frequency domain such that the pairing problem can be carried out over a range of frequencies that are of practical significance for the plant. It also provides dynamic measures for interaction and sensitivity. In

A. Khaki-Sedigh and B. Moaveni: Control Configuration Selection, LNCIS 391, pp. 99–113.
springerlink.com

(Grosdidier and Morari 1986), structured singular values are used to provide another interaction measure and a unified treatment is applied to the known interaction measures up to that time. This IM is employed to study the design of stabilizing decentralized controllers through stabilizing the block diagonal part of the plant.

Consider the multivariable closed loop plant shown in figure 4.1. Let $G(s)$ be an $m \times m$ transfer function matrix and $C(s) = diag\{C_1(s) \;\; \cdots \;\; C_M(s)\}$ be the block decentralized controller, where $C_i(s)$ are $m_i \times m_i$ block controllers and $i = 1, 2, \cdots, M$; $\sum_{i=1}^{M} m_i = m$. The plant transfer function matrix, reference input, control signal and output vectors shown in figure 4.1 are partitioned in the same manner. It is assumed that the block diagonal closed loop plant is designed stable, that is

$$\tilde{T}(s) = \tilde{G}(s)C(s)\left[I + \tilde{G}(s)C(s)\right]^{-1} \tag{4.1}$$

is stable. However, the full closed loop plant is

$$T(s) = G(s)C(s)\left[I + G(s)C(s)\right]^{-1} \tag{4.2}$$

and an IM will be defined to derive the conditions to ensure the stability of $T(s)$. We now define the following *relative error matrix*

$$E(s) = \left[G(s) - \tilde{G}(s)\right]\tilde{G}^{-1}(s) \tag{4.3}$$

note that $G(s) - \tilde{G}(s)$ is the *error* caused by approximating the full plant by its equivalent block diagonal part.

Denote the net number of clockwise encirclements of the point $(k, 0)$ by the Nyquist plot of $g(s)$ as $N\left(k, g(s)\right)$. Then, the following key result on the stability of $T(s)$ is proved in (Grosdidier and Morari 1986), and with a minor modification in (Kariwala et al. 2007).

Theorem 4.1

Assume that $G(s)$ and $\tilde{G}(s)$ have the same number of RHP poles and that $\tilde{T}(s)$ is stable. Then, $T(s)$ is stable if and only if

$$\det\left[I + E(s)\tilde{T}(s)\right] \neq 0 \tag{4.4}$$

$$N\left(0, \det\left[I + E(s)\tilde{T}(s)\right]\right) = 0 \tag{4.5}$$

■

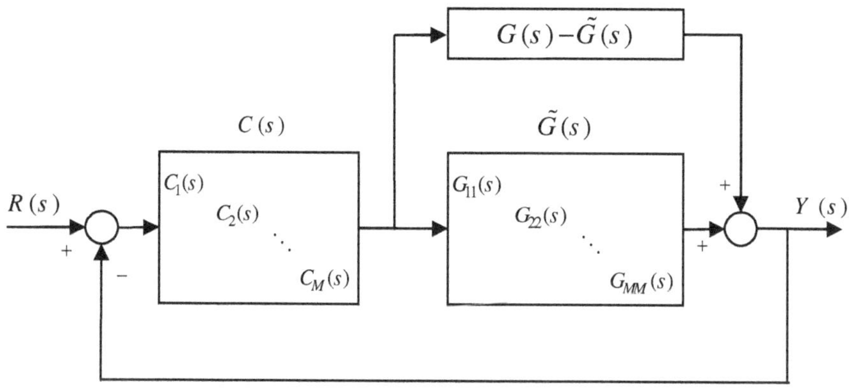

Fig. 4.1 Block diagonal representation of interaction as additive uncertainty.

Note 1. The assumption that $G(s)$ and $\tilde{G}(s)$ have the same number of RHP poles is only valid if the location and multiplicity of the RHP pole is the same or the transfer function matrix has a special structure. For example, let

$$G(s) = \frac{1}{(s-1)(s+1)}\begin{bmatrix} 2(s-1) & (s-1) \\ 1 & 1 \end{bmatrix}$$

Then, it is easily verified that $G(s)$ and $\tilde{G}(s)$ have the same unstable pole at 1. It is obvious that the assumption is trivially valid for the open loop stable multivariable plants.

Note 2. For diagonal $G(s)$, $E(s)$ is the zero matrix and equation (4.5) is trivially satisfied.

Note 3. For non diagonal $G(s)$, the size of the $E(s)$ matrix depends on the interaction in the plant. It is intuitively argued that to avoid encirclements and hence satisfying equation (4.5) for *large* $E(s)$, $\tilde{T}(s)$ must be *small*. This would result in closed loop performance degradation and this shows the necessary compromise between closed loop stability and performance.

In (Grosdidier and Morari 1986), a generalization of the NI (see section 2.2.3) to block diagonal controllers is given. First assume that $G(s)$ and $\tilde{G}(s)$ have the same RHP poles and $\tilde{T}(s)$ is designed stable. Also, let $G(s)C(s)$ be strictly proper. This implies that

$$\lim_{s \to \infty} \det\left(I + E(s)\tilde{T}(s)\right) = 1 \tag{4.6}$$

Finally, assume that steady state set point tracking is achieved, i.e. $\tilde{T}(0) = I$. This implies that

$$\lim_{s \to 0} \det\left(I + E(s)\tilde{T}(s)\right) = \lim_{s \to 0} \det\left(G(s)\tilde{G}^{-1}(s)\right) \tag{4.7}$$

It is now obvious that equation (4.5) will not be satisfied if $\lim_{s \to 0} \det\left(G(s)\tilde{G}^{-1}(s)\right) < 0$, since this requires the Nyquist plot to start from the left and equation (4.6) requires the Nyquist plot to end at right of the origin. Therefore, the closed loop plant $T(s)$ is unstable if

$$\lim_{s \to 0} \det\left(G(s)\tilde{G}^{-1}(s)\right) < 0 \tag{4.8}$$

Inequality (4.8) provides the generalization of NI to block diagonal controllers and open loop unstable plants. However, note that it only gives a necessary stability condition. Inequality (4.8) can alternatively be written as

$$\frac{\det\left(G(0)\right)}{\prod_{i=1}^{m} \det\left(G_{ii}(0)\right)} < 0 \tag{4.9}$$

Note. Inequality (4.8) is not satisfied for diagonal $G(s)$. However, if $G(s)$ and $\tilde{G}(s)$ are *very different*, which implies that the plant has high interactions, inequality (4.8) may hold. In such cases, $\tilde{T}(0) = I$ will lead to closed loop instability. Hence, we should impose the constraints $\tilde{T}(0) \neq I$ to avoid the closed loop instability. But, this constraint would mean that the closed loop plant will have steady state error. This results in a compromise between stability and performance.

Before presenting the main result of this section, some definitions are given. Let $\bar{\sigma}(A)$ denote the maximum singular value of the $m \times n$ matrix A, and define $\Delta = \left\{diag\left(\Delta_i\right) : \Delta_i \text{ are } m_i \times n_i \text{ matrices}\right\}$. Then, the structured singular value (SSV) of A is given as

$$\mu_\Delta(A) = \frac{1}{\min\left\{\bar{\sigma}(\tilde{\Delta}) : \tilde{\Delta} \in \Delta, \det(I - A\tilde{\Delta}) = 0\right\}}$$

and $\mu_\Delta(A)=0$ if there is no $\tilde{\Delta}\in\Delta$ such that $\det(I-A\tilde{\Delta})=0$. We now give the $\mu-$IM based on the SSV definition.

Theorem 4.2

Assume that $G(s)$ and $\tilde{G}(s)$ have the same number of RHP poles and that $\tilde{T}(s)$ is stable. Then, $T(s)$ is stable if

$$\bar{\sigma}(\tilde{T}(j\omega))<\mu_\Delta^{-1}\left(E(j\omega)\right) \qquad \forall\omega \tag{4.10}$$

where Δ has the same block structure as $\tilde{G}(s)$.

■

Equation (4.10) is called the $\mu-$IM. This provides the condition that gives closed loop stability when the design is performed using the block diagonal part of the transfer function matrix only.

Note. In the case of integral control, $\mu_\Delta^{-1}\left(E(0)\right)>1$ is a sufficient condition for DIC. This condition can be used to ensure the DIC of a particular pairing. However, it can not be used to eliminate a pairing that does not satisfy this inequality (Skogestad and Morari 1992).

Example 4.2.1

Consider the distillation column of Doukas and Luyben with 3×3 transfer function matrix as (Grosdidier and Morari 1986)

$$G(s)=\begin{bmatrix} \dfrac{0.374e^{-7.75s}}{(22.2s+1)^2} & \dfrac{-11.3e^{-3.79s}}{(21.74s+1)^2} & \dfrac{-9.811e^{-1.59s}}{(11.36s+1)} \\ \dfrac{-1.986e^{-0.71s}}{(66.67s+1)^2} & \dfrac{5.24e^{-60s}}{(400s+1)} & \dfrac{5.984e^{-2.24s}}{(14.29s+1)} \\ \dfrac{0.0204e^{-0.59s}}{(7.14s+1)^2} & \dfrac{-0.33e^{-0.68s}}{(2.38s+1)^2} & \dfrac{2.38e^{-0.42s}}{(1.43s+1)^2} \end{bmatrix}$$

The corresponding RGA is

$$\Lambda=\begin{bmatrix} -0.0986 & 1.0004 & 0.0983 \\ 1.0926 & -0.1043 & 0.0117 \\ 0.0060 & 0.1093 & 0.8900 \end{bmatrix}$$

and it is obvious that diagonal pairing is not appropriate. In figure 4.2, $\mu_\Delta^{-1}(E(j\omega))$ for diagonal pairing $(u_1-y_1,u_2-y_2,u_3-y_3)$, and block diagonal pairing $(u_{1,2}-y_{1,2},u_3-y_3)$ are shown. On the basis of theorem 4.2 diagonal pairing may not result in a stable decentralized control with integral action, while the block diagonal pairing can stabilize the closed loop plant, as $\mu_\Delta^{-1}(E(0))>1$. Note that $T(0)=I$ for tracking with integral action, and the inequality (4.10) is clearly violated by the diagonal pairing.

Next, we use the $\mu-$IM for $(u_1-y_2,u_2-y_1,u_3-y_3)$ and $(u_1-y_3,u_2-y_1,u_3-y_2)$. $\mu_\Delta^{-1}(E(j\omega))$ for these two pairings are shown in figure 4.3 and it is obvious that $(u_1-y_2,u_2-y_1,u_3-y_3)$ is the appropriate input-output pair. It is interesting to note that the RGA analysis and $\mu-$IM give similar results. ∎

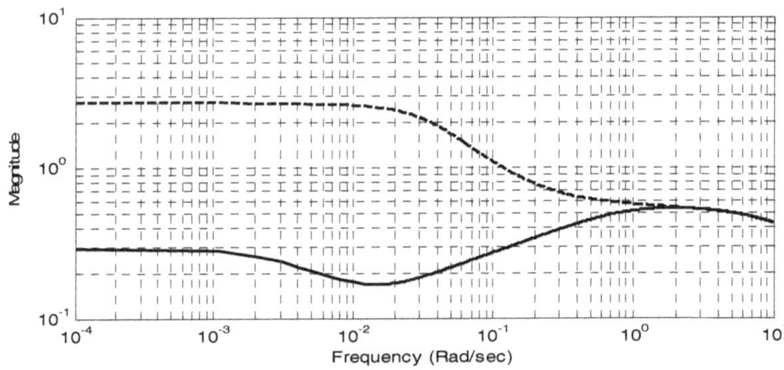

Fig. 4.2 $\mu_\Delta^{-1}(E(j\omega))$ frequency response. Solid line corresponds to the diagonal pairing and dashed line corresponds to the block diagonal pairing $(u_{1,2}-y_{1,2},u_3-y_3)$.

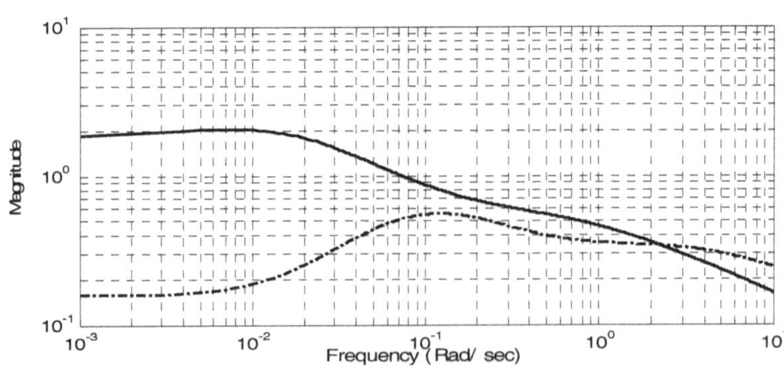

Fig. 4.3 Frequency response of $\mu_\Delta^{-1}(E(j\omega))$. Solid line corresponds to $(u_1-y_2,u_2-y_1,u_3-y_3)$ and Dash-dot line corresponds to $(u_1-y_3,u_2-y_1,u_3-y_2)$.

Example 4.2.2

Consider the multivariable plant of example 2.3.3 with the following transfer function matrix

$$G(s)=\begin{bmatrix}\dfrac{5e^{-40s}}{100s+1} & \dfrac{e^{-4s}}{10s+1}\\ \dfrac{-5e^{-4s}}{10s+1} & \dfrac{5e^{-40s}}{100s+1}\end{bmatrix}$$

where it is shown in example 2.3.3 that the corresponding RGA proposes the diagonal pairing. In figure 4.4, $\mu_\Delta^{-1}(E(j\omega))$ for diagonal and off-diagonal pairings are shown. According to the theorem 4.2, diagonal pairing (u_1-y_1,u_2-y_2) is the recommended input-output pair.

■

4.3 An Alternative Approach to Control Configuration Selection Based on Passivity

In a rather different but promising approach to the control configuration selection, the concept of passivity has been recently adopted. This section presents the passivity based pairing methodology and follows the derivations in (Bao et al. 2007).

4.3.1 Passivity Definition and Fundamental Results

The properties and applications of passivity theory and passive systems are well known in stability analysis and control systems design. However, the concept has

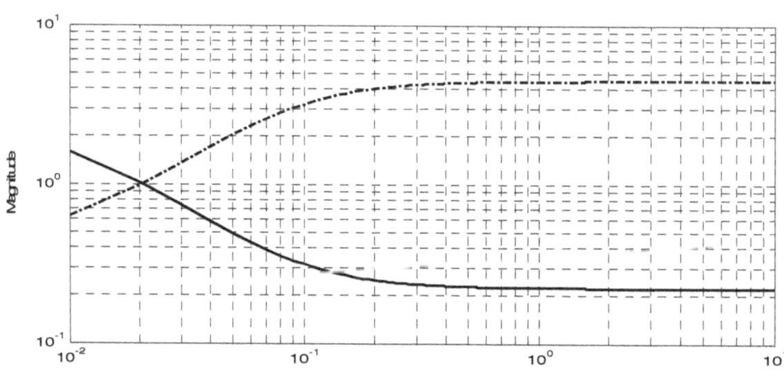

Fig. 4.4 Frequency response of $\mu_\Delta^{-1}(E(j\omega))$. Solid line corresponds to (u_1-y_1,u_2-y_2) and Dash-dot line corresponds to (u_1-y_2,u_2-y_1).

been newly introduced in interaction analysis (Bao et al. 2000), (Bao et al. 2002) and (Bao et al. 2007). Contrasting the conventional pairing methods that use the generalized diagonal dominance, the proposed method introduces a frequency dependent passivity index to pair the systems according to their so-called *degree of passivity*. It is claimed that this index provides the best achievable closed loop performance under decentralized control. Experimental nature and the simple open loop step response tests required for the implementation of the method are the main advantages of the proposed methodology. These concepts are described using the results of (Bao et al. 2007). Let, $H(z)$ be the pulse transfer function matrix of the plant.

Definition 4.1. A square multivariable discrete time plant with rational function elements analytic outside the unit circle is *passive* if and only if:

- Poles of the elements of $H(z)$ on the unit circle are simple.
- $H(e^{j\theta})+H^T(e^{-j\theta})\geq 0$ for all real θ at which $H(e^{j\theta})$ exists.
- The matrix $Q=e^{j\theta_0}R$ is a semi-positive definite Hermitian matrix, where $e^{j\theta_0}$ is a pole of an element of $H(z)$ and R is its corresponding residue matrix.

Moreover, it is *strictly passive* if it is asymptotically stable and $H(e^{j\theta})+H^T(e^{-j\theta})>0$ for all real θ.

Note. $\theta = 2\pi\omega/\omega_s$, where ω_s is the sampling frequency and $H(e^{j\theta})$ is the frequency response of the plant $H(z)$.

The passivity condition does not imply diagonal dominance and vice versa. The following example shows this point.

Example 4.3.1

Consider the following transfer function matrix

$$H(z)=\begin{bmatrix} \dfrac{z}{2z-1} & -a \\ a & \dfrac{z}{3z-1} \end{bmatrix} \qquad a\in\Re$$

It can be easily shown that $H(z)$ is passive for all a, but it is not diagonally dominant for large enough a.

■

Consider the multivariable closed loop plant shown in figure 4.5. Passivity theorem states that for a strictly passive multivariable plant $H(z)$, regardless of the amount of interactions between different loops, closed loop stabilization is possible by any passive multi-loop controller $K(z)$. However, if the plant is not passive there will be performance limitations, achievable by passive controllers. Hence, the degree of passivity of the plant with a particular control configuration can show the achievable performance of that configuration. The analysis in (Bao et al. 2007) demonstrates that the *extent* of passivity for discrete time plants with a selected control configuration is quantified by a passivity index. Hence, a suitable input-output pair is the pair that leads to the *most* passive plant model subject to certain controller output gain constraints.

A key property is the fact that stable multivariable plants can be made passive by a dynamic feedforward compensator as shown in the following theorem proved in (Bao et al. 2007).

Theorem 4.3

For a given stable non passive multivariable plant $H(z)$ there exists a diagonal, stable and passive transfer function matrix $W(z)=w(z)I$ such that $H'(z)=H(z)+W(z)$ is passive.

■

Note. Unstable multivariable plants with minimum phase zeros can always be rendered passive by feedback compensation.

Definition 4.2. Let $H(z)$ be a square stable multivariable discrete time plant. The frequency dependent *passivity index* is defined as

$$v\_(H(z),\theta) \triangleq \max\left\{-\frac{1}{2}\lambda_{\min}\left[H(e^{j\theta})+H^T(e^{-j\theta})\right],0\right\} \tag{4.11}$$

where $\lambda_{\min}[\cdot]$ is the minimum eigenvalue and $0 \le \theta \le \pi$.

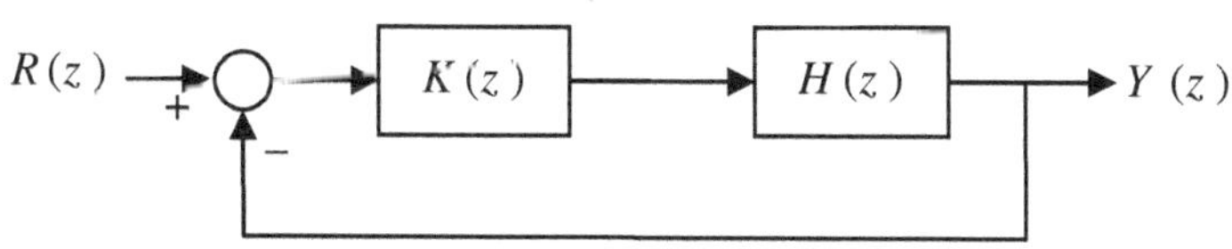

Fig. 4.5 Multi-loop control system.

Note. The above passivity index corresponds to diagonal elements. Hence, to compute the passivity index for other pairs the plant should be reordered.

This index is a measure to quantify the required amount of compensation to passify the plant. A positive value of the passivity index indicates the lack of passivity.

Note that positive passivity index requires negative eigenvalues of the matrix $H(e^{j\theta})+H^T(e^{-j\theta})$ which gives a non positive matrix. It is shown that the phase of a multivariable passive plant $H(z)$ and each control loop in a multivariable decentralized passive control system lies between $-\pi/2$ and $\pi/2$ (Bao et al. 1998). To use a passive controller, the diagonal elements of $H(z)$ must be non negative at steady state. Hence, $H(z)$ is rescaled as follows

$$H^+(z)=H(z)\Upsilon \tag{4.12}$$

where $\Upsilon=diag\{\upsilon_i\}$ for $i=1,\cdots,m$ and $\upsilon_i=\pm 1$ such that $\left[H^+(1)\right]_{ii}\geq 0$ for $i=1,\cdots,m$.

Two separate cases may occur in practice. First, if $H^+(z)$ is passive the passive controller $K^+(z)=diag\left\{k_i^+(z)\right\}$ can be designed and the final controller is $K(z)=\Upsilon K^+(z)$. However, if $H^+(z)$ is not passive but assumed stable, the plant is first passified shown by $H'(z)$ and the stabilizing controller denoted by $K'(z)$ can be designed for the passified plant. This is shown in figure 4.6. Now the final controller $K(z)$ can be derived by rearranging the loop shown in figure 4.6. This gives $K(z)=\Upsilon K'(z)\left[I+w(z)K'(z)\right]^{-1}$. The following theorem proved in (Bao et al. 2007) presents the main closed loop stability result.

Theorem 4.4

The closed loop plant shown in figure 4.6 is stable if the controller $K(z)$ does not posses any poles outside the unit circle and for all $\theta\in[0,\pi]$ and $i=1,\cdots,m$ the following condition hold

$$\mathrm{Re}\left[\frac{k_i^+(e^{j\theta})}{1-w(e^{j\theta})k_i^+(e^{j\theta})}\right]>0 \tag{4.13}$$

where $\mathrm{Re}\left[w(e^{j\theta})\right]>v\_(H^+(z),\theta)$.

■

Note 1. It is interesting to note that if $k_i'(z)$ is passive then $K'(z)$ will remain passive for $K'(z)E = diag\{k_i'\varepsilon_i\}$, $0 \le \varepsilon_i \le 1$ and $i = 1, \cdots, m$, hence ensuring the decentralized unconditional stability of the closed loop plant.

Note 2. If the passivity index of $H(z)$ at steady state is zero, the decentralized controllers can have infinite steady state gain. Also, there would be at least one decentralized controller with integral action that stabilizes the plant. That is zero steady state error with guaranteed closed loop stability is achieved in the face of detuning one or more controllers arbitrarily. Hence, a sufficient condition of DIC is provided.

Note 3. To facilitate the applicability of the method, it is shown in (Bao et al. 2007) that this index can be derived directly from open loop step responses.

4.3.2 Passivity Based Pairing Rules

The main pairing rules based on the passivity concept are summarized as in (Bao et al. 2007). As different pairing schemes result in different transfer function matrices $H(z)$ with normally different passivity indices, the selected pairing should give a *more* passive $H(z)$. The controller can have an infinite steady state gain, which results in high control performance in the frequency band for which $H(z)$ is passive. Hence, the first pairing rule is:

- Select the pairing that results in a passive $H(z)$ with the largest frequency band.

Although high control gains lead to good closed loop performance, but the actuator and controller output constraints is a serious limitation in selecting the pairing scheme. Consider the closed loop plant shown in figure 4.5. The transfer function matrix from the reference input to the controller output is

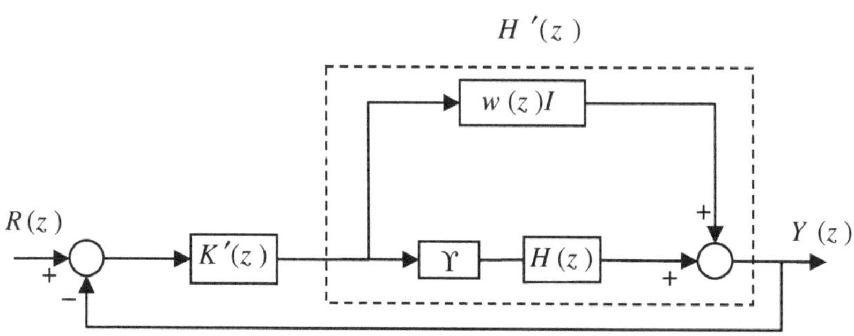

Fig. 4.6 Passified system $H'(z)$ with passive controller $K'(z)$.

$$T_c(z) = K(z)\left[I + H(z)K(z)\right]^{-1} \tag{4.14}$$

In the case of integral action in the control loops, we have $K(1) = \infty$ which gives $T_c(1) = H^{-1}(1)$. Hence, the steady state relationship between the m dimensional reference input and the controller output is $U = H^{-1}(1)R$. To limit the controller output for implementation purposes, the pairing scheme should lead to a sufficiently small $H^{-1}(1)$. This would give a high performance closed loop control system without violating the existing practical limit constraints. Let these limits be shown as $|u_i| \le \alpha_i$ and $|r_i| \le \beta_i$ $(i = 1,\cdots,m)$ for the controller outputs and the set points, respectively. Hence, ignoring the interactions, we have the following necessary condition

$$\left|\left[H^{-1}(1)\right]_{ii}\right| \le \frac{\alpha_i}{\beta_i} \qquad i = 1,\ldots,m \tag{4.15}$$

which gives the second pairing rule as follows:

- To fulfill the controller output constraints in practical applications, the pairing scheme should satisfy inequality (4.15).

Example 4.3.1

Consider the following 2×2 transfer function matrix

$$G(s) = \begin{bmatrix} \dfrac{-2e^{-s}}{10s+1} & \dfrac{1.5e^{-s}}{s+1} \\ \dfrac{1.5e^{-s}}{s+1} & \dfrac{-2e^{-s}}{10s+1} \end{bmatrix}$$

where the corresponding RGA is given below and it proposes the diagonal pairing as the appropriate input-output pair

$$\Lambda = \begin{bmatrix} 2.2857 & -1.2857 \\ -1.2857 & 2.2857 \end{bmatrix}$$

The passivity indices versus the frequency are calculated and shown in figure 4.7. It is obvious that the off-diagonal pair has positive passivity index at $\theta = 0$. Hence, the off-diagonal pairing is not DIC. So, the passivity based approach proposes the $(u_1 - y_1, u_2 - y_2)$ as the appropriate input-output pair, which is compatible with the result of the RGA analysis.

■

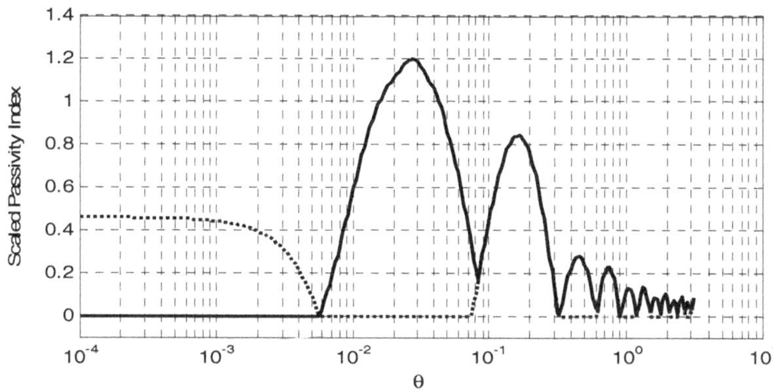

Fig. 4.7 Frequency responses of passivity indices. Solid line corresponds to $(u_1 - y_1, u_2 - y_2)$ and dotted line corresponds to $(u_1 - y_2, u_2 - y_1)$.

Example 4.3.2

Consider a 3×3 distillation column with continuous transfer function matrix as

$$G(s) = \begin{bmatrix} \dfrac{-1.986e^{-0.71s}}{66.67s+1} & \dfrac{5.24e^{-60s}}{400s+1} & \dfrac{5.984e^{-2.24s}}{14.29s+1} \\ \dfrac{0.0204e^{-4.199s}}{5s+1} & \dfrac{-0.33e^{-1.883s}}{3.904s+1} & \dfrac{2.38e^{-1.143s}}{10s+1} \\ \dfrac{0.374e^{-7.75s}}{22.22s+1} & \dfrac{-11.3e^{-14.78s}}{35.66s+1} & \dfrac{-9.881e^{-1.59s}}{11.35s+1} \end{bmatrix}$$

with constraints on the control effort and set points given as $|u_1|\le 1=\alpha_1$, $|u_2|\le 0.1=\alpha_2$, $|u_3|\le 0.5=\alpha_3$ and $|r_i|\le 1=\beta_i$ for $i=1,2,3$. Let the sampling period be 1 minute and the settling time 2400 minutes. Input-output pairing analysis in all the 6 cases shown in table 4.1 should be studied. It is observed that only $(u_1 - y_1, u_2 - y_3, u_3 - y_2)$ and $(u_1 - y_3, u_2 - y_1, u_3 - y_2)$ satisfy the DIC and constraints conditions. Hence, the appropriate input-output pair is the one with the largest frequency bandwidth $[0, \theta_b]$, where $v\_(H^+(z), \theta) = 0$ for $\theta \in [0, \theta_b]$, according to the first paring rule. Figure 4.8 shows the frequency responses of passivity indices for $(u_1 - y_1, u_2 - y_3, u_3 - y_2)$ and $(u_1 - y_3, u_2 - y_1, u_3 - y_2)$. It is obvious that $(u_1 - y_1, u_2 - y_3, u_3 - y_2)$ has the larger frequency bandwidth, $\theta_{b1} > \theta_{b2}$ and is therefore the proposed input-output pair.

■

Table 4.1 Possible input-output pairs in example 4.3.2

	Input-Output Pair	DIC condition $\nu\_(H^{+}(z),0)=0$	$\left\| \left[H^{-1}(1)\right]_{ii} \right\| \leq \alpha_i / \beta_i$
1	$u_1-y_1, u_2-y_2, u_3-y_3$	**OK**	**Failed**
2	$u_1-y_1, u_2-y_3, u_3-y_2$	**OK**	**OK**
3	$u_1-y_3, u_2-y_2, u_3-y_1$	**Failed**	**Failed**
4	$u_1-y_2, u_2-y_1, u_3-y_3$	**Failed**	**OK**
5	$u_1-y_3, u_2-y_1, u_3-y_2$	**OK**	**OK**
6	$u_1-y_2, u_2-y_3, u_3-y_1$	**Failed**	**OK**

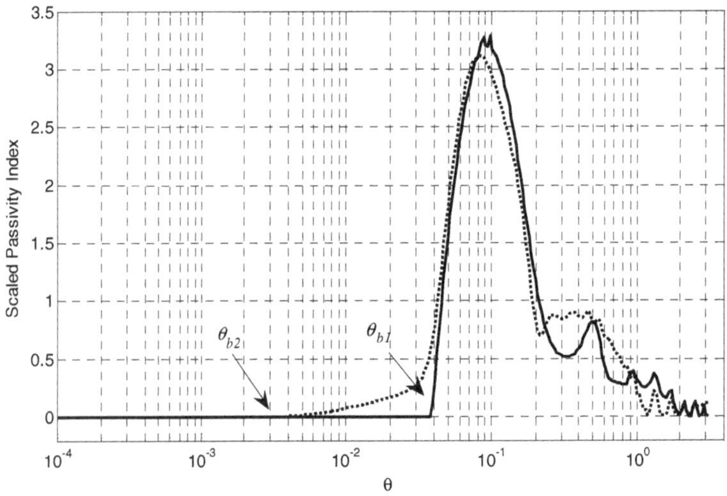

Fig. 4.8 Frequency responses of passivity indices. Solid line corresponds to $(u_1-y_1, u_2-y_3, u_3-y_2)$ and dotted line corresponds to $(u_1-y_3, u_2-y_1, u_3-y_2)$.

4.4 Conclusion

The main purpose of this chapter has been to provide an overview of the advanced pairing techniques which are not direct generalizations of the RGA. These include the SSV and passivity based techniques. Examples are used to show the main points of the mentioned pairing methodologies.

References

Bao, J., Chan, K.H., Zhang, W.Z., Lee, P.L.: An experimental pairing method for multi-loop control based on passivity. J. Process Contr. 17, 787–798 (2007)

Bao, J., Lee, P.L., Wang, F.Y., Zhou, W.B.: New robust stability criterion and robust controller synthesis. Int. J. Robust Nonlin. 8, 49–59 (1998)

Bao, J., Lee, P.L., Wang, F.Y., Zhou, W.B., Samyudia, Y.: A new approach to decentralized process control using passivity and sector stability conditions. Chem. Eng. Commun. 182, 213–237 (2000)

Bao, J., McLellan, P.J., Forbes, J.F.: A passivity-based analysis for decentralized integral controllability. Automatica 38, 243–247 (2002)

Grosdidier, P., Morari, M.: Interaction measures for systems under decentralized control. Automatica 22, 309–319 (1986)

Kariwala, V., Forbes, J.F., Skogestad, S.: [graphics object to be inserted manually]-Interaction measure for unstable systems. Int. J. Automation and Control 1, 295–313 (2007)

Lau, H., Alvarez, J., Jensen, K.F.: Synthesis of control structures by singular value analysis: dynamic measures of sensitivity and interaction. AIChE J. 31, 427–439 (1985)

Morari, M.: Operability measures for process design. Ind. Chem. Eng. Symposium Series, 131–140 (1982)

Skogestad, S., Morari, M.: Variable selection for decentralized control. Modeling, Identification, and Control 13, 113–125 (1992)

Smith, C.R., Moore, C.F., Bruns, D.D.: A structural framework for multivariable control applications. In: Proceeding of the Joint American Control Conference, Charlottesville, Virginia (1981)

Smith, C.R.: Multivariable process control using SVD. PhD Thesis, University of Tennessee, USA (1981)

Chapter 5
Control Configuration Selection of Linear Multivariable Plants Based on the State-Space Models

5.1 Introduction

In the previous chapters, input-output pairing methods based on the transfer function model of the plant are introduced. The conventional RGA uses the steady state transfer function matrix, and the advanced pairing methods based on the RGA concept or the passivity approach use the dynamic transfer function matrix. In this chapter, input-output pairing methods based on the dynamical state-space model of the plant are introduced. State space is an internal description of the plant and it is therefore expected to support effective control structure methodologies.

In section 5.2, a brief introduction to necessary definitions and results of linear system theory used in the subsequent sections, are presented. Section 5.3 introduces the input-output pairing method based on the singular perturbation methodology for linear multivariable plants with significant different time constants. In sections 5.4, 5.5 and 5.6 three input-output pairing methods, using the Gramian matrices are presented. In section 5.4, the Participation Matrix is introduced that uses the summation of the Hankel singular values to measure the interaction. Also, this method introduces an overall interaction measure to choose the best input-output pair among the appropriate pairs. Section 5.5 introduces the Hankel Interaction Index Array to control structure selection. In the last section, Dynamic Input-Output Pairing Matrix based on the Hankel norm of elementary subsystems is introduced.

5.2 Elements of Linear System Theory

5.2.1 Controllability and Observability Gramians

Consider the linear single-input single-output (SISO), asymptotically stable, time invariant plant $S(A_{n\times n}, b_{n\times 1}, c_{1\times n})$ described by the following state space equations

A. Khaki-Sedigh and B. Moaveni: Control Configuration Selection, LNCIS 391, pp. 115–138.
springerlink.com

$$\dot{x}(t) = Ax(t) + bu(t)$$
$$y(t) = cx(t) \tag{5.1}$$

The controllability and observability Gramian matrices are defined as

$$W_c \triangleq \int_0^\infty e^{At} bb^T e^{A^T t} dt \tag{5.2}$$

and

$$W_o \triangleq \int_0^\infty e^{A^T t} c^T c e^{At} dt \tag{5.3}$$

respectively. These matrices can be computed by solving the following linear Lyapunov equations

$$W_c A^T + A W_c = -bb^T \tag{5.4}$$

and

$$W_o A + A^T W_o = -c^T c \tag{5.5}$$

The plant described by equations (5.1) is controllable and observable if W_c and W_o are nonsingular matrices (Fernando and Nicholson 1983). Gramian matrices depend on the state space realization, but the eigenvalues of their product, λ_i $(i = 1, 2, \ldots, n)$, are nonnegative and are independent of the state space realization. The Hankel Singular Values (HSV) are defined as

$$\sigma_H^{(i)} = \sqrt{\lambda_i}, \qquad i = 1, 2, \ldots, n \tag{5.6}$$

where $\sigma_H^{(1)} \geq \sigma_H^{(2)} \geq \cdots \geq \sigma_H^{(l)} > 0$ are the nonzero HSV. HSV provide some valuable information regarding the plant. A selected summary of this information is given below:

- The number of nonzero HSV l, shows the dimension of controllable and observable subspace.
- The ratio of the maximum and minimum (nonzero) HSV, $\sigma_H^{(1)}/\sigma_H^{(l)}$, is a measure of the skewness of the controllability and observability of the controllable and observable subspaces.
- Maximum HSV of the plant is the system Hankel norm denoted by $\|G(s)\|_H = \|S(A,b,c)\|_H$.

An interesting property of the Gramians is their relation with the plant 2-norm presented in (Doyle et al. 1992) as

$$\|G(s)\|_2^2 = \mathrm{trace}(cW_c c^T) = \mathrm{trace}(b^T W_o b) \tag{5.7}$$

The cross-Gramian matrix is another attractive matrix which contains valuable information about the plant controllability and observability. Using the impulse response of the controllable and observable plant, the cross-Gramian matrix W_{co} is defined as (Fernando and Nicholson 1983)

$$W_{co} \triangleq \int_0^\infty (e^{At}b)(e^{A^Tt}c^T)^T\,dt = \int_0^\infty e^{At}bce^{At}dt \tag{5.8}$$

It is easily seen that the matrix W_{co} can be computed by solving the following Sylvester equation

$$W_{co}A + AW_{co} = -bc \tag{5.9}$$

Since the matrix A is assumed stable, a unique solution matrix W_{co} exists. It is intuitively clear that the matrix W_{co} carries information about both controllability and observability properties and it is easily shown that the eigenvalues of W_{co} are invariant under similarity transformations.

Also, in (Moaveni and Khaki Sedigh 2008) a simple method to solve the Sylvester equation and to compute the cross-Gramian matrix is introduced. Two approaches for distinct and repeated eigenvalues are provided to give the solution for a class of Sylvester equations given by equation (5.9).

Consider the plant described by equations (5.1). Let λ_i ($i = 1,\ldots,n$) be distinct eigenvalues of the matrix A and v_i ($i = 1,\ldots,n$) be the corresponding eigenvectors. The cross-Gramian matrix W_{co} can be computed using the Sylvester equation (5.9). Multiplying the Sylvester equation by eigenvector v_i gives

$$AW_{co}v_i + W_{co}Av_i = -bcv_i \tag{5.10}$$

and

$$(A + \lambda_i I)W_{co}v_i = -bcv_i \tag{5.11}$$

and

$$W_{co}v_i = -(A+\lambda_i I)^{-1}bcv_i, \qquad (i=1,\ldots,n) \tag{5.12}$$

Rewriting these n equations in matrix form gives

$$W_{co}\begin{bmatrix} v_1 & \cdots & v_n \end{bmatrix} = -\begin{bmatrix} (A+\lambda_1 I)^{-1}bcv_1 & \cdots & (A+\lambda_n I)^{-1}bcv_n \end{bmatrix} \tag{5.13}$$

Hence, the cross-Gramian matrix can be computed as

$$W_{co} = -\begin{bmatrix} I \\ \vdots \\ I \\ \vdots \\ I \end{bmatrix}^T \begin{bmatrix} (A+\lambda_1 I)^{-1}bcv_1 & 0 & \cdots & \cdots & 0 \\ 0 & \ddots & 0 & \ddots & \vdots \\ \vdots & 0 & (A+\lambda_i I)^{-1}bcv_i & 0 & \vdots \\ \vdots & \ddots & 0 & \ddots & 0 \\ 0 & \cdots & \cdots & 0 & (A+\lambda_n I)^{-1}bcv_n \end{bmatrix} \begin{bmatrix} v_1^T \\ \vdots \\ v_i^T \\ \vdots \\ v_n^T \end{bmatrix}^{-T} \tag{5.14}$$

where I is the $n \times n$ identity matrix and 0 is a $n \times 1$ zero vector.

Consider the plant with repeated eigenvalue λ and multiplicity n. Let, $v = v^{(1)}$ be the eigenvector and $v^{(i)}$ ($i = 2,\ldots,n$) be the corresponding generalized eigenvectors. Then, the corresponding Jordan form is

$$J = \begin{bmatrix} \lambda & 1 & 0 & \cdots & 0 \\ 0 & \lambda & 1 & & \vdots \\ \vdots & 0 & \ddots & \ddots & 0 \\ \vdots & \vdots & 0 & \ddots & 1 \\ 0 & 0 & \cdots & 0 & \lambda \end{bmatrix} \tag{5.15}$$

the Sylvester equation (5.9) gives

$$AW_{co}v^{(1)} + W_{co}Av^{(1)} = -bcv^{(1)} \tag{5.16}$$

using equations (5.11) and (5.12), equation (5.16) can be rewritten as

$$W_{co}v^{(1)} = -(A+\lambda I)^{-1}bcv^{(1)} \tag{5.17}$$

Also, for generalized eigenvectors equation (5.9) gives

$$AW_{co}v^{(2)} + W_{co}Av^{(2)} = -bcv^{(2)} \tag{5.18}$$

and it is well known that

$$(A-\lambda I)v^{(2)}=v^{(1)} \tag{5.19}$$

Then, equation (5.18) can be rewritten as

$$AW_{co}v^{(2)}+\lambda W_{co}v^{(2)}+W_{co}v^{(1)}=-bcv^{(2)} \tag{5.20}$$

Using equation (5.17), equation (5.20) gives

$$AW_{co}v^{(2)}+\lambda W_{co}v^{(2)}=(A+\lambda I)^{-1}bcv^{(1)}-bcv^{(2)} \tag{5.21}$$

and

$$W_{co}v^{(2)}=(A+\lambda I)^{-2}bcv^{(1)}-(A+\lambda I)^{-1}bcv^{(2)} \tag{5.22}$$

Similarly for $i=1,\ldots,n$

$$\begin{aligned}W_{co}v^{(i)}=&(-1)^{i}(A+\lambda I)^{-i}bcv^{(1)}+(-1)^{i-1}(A+\lambda I)^{-i+1}bcv^{(2)}+\cdots\\&+(A+\lambda I)^{-2}bcv^{(i-1)}-(A+\lambda I)^{-1}bcv^{(i)}\end{aligned} \tag{5.23}$$

Rewriting these equations in matrix form gives

$$W_{co}\begin{bmatrix}v^{(1)T}\\v^{(2)T}\\ \\v^{(n)T}\end{bmatrix}^{-T}=\begin{bmatrix}I\\I\\\vdots\\I\end{bmatrix}^{T}\begin{bmatrix}-(A+\lambda I)^{-1}bcv^{(1)} & -(A+\lambda I)^{-1}bcv^{(2)} & \cdots & -(A+\lambda I)^{-1}bcv^{(n)}\\0 & (A+\lambda I)^{-2}bcv^{(1)} & \ddots & \vdots\\\vdots & 0 & \ddots & (-1)^{n-1}(A+\lambda I)^{-(n-1)}bcv^{(2)}\\0 & \cdots & 0 & (-1)^{n}(A+\lambda I)^{-n}bcv^{(1)}\end{bmatrix} \tag{5.24}$$

Hence, the cross Gramian matrix can be computed as

$$W_{co}=\begin{bmatrix}I\\I\\\vdots\\I\end{bmatrix}^{T}\begin{bmatrix}-(A+\lambda I)^{-1}bcv^{(1)} & -(A+\lambda I)^{-1}bcv^{(2)} & \cdots & -(A+\lambda I)^{-1}bcv^{(n)}\\0 & (A+\lambda I)^{-2}bcv^{(1)} & \ddots & \vdots\\\vdots & 0 & \ddots & (-1)^{n-1}(A+\lambda I)^{-(n-1)}bcv^{(2)}\\0 & \cdots & 0 & (-1)^{n}(A+\lambda I)^{-n}bcv^{(1)}\end{bmatrix}\begin{bmatrix}v^{(1)T}\\v^{(2)T}\\ \\v^{(n)T}\end{bmatrix}^{-T} \tag{5.25}$$

where I is the $n\times n$ identity matrix and 0 is a $n\times 1$ zero vector.

5.2.2 Balanced Realization

Consider the plant described by $S(A_{n\times n},b_{n\times 1},c_{1\times n})$. Employing the similarity transformation matrix T, the transformed state space description $S(A^{b}_{n\times n},b^{b}_{n\times 1},c^{b}_{1\times n})$ is

$$\begin{aligned} A^b &= T^{-1}AT \\ b^b &= T^{-1}b \\ c^b &= cT \end{aligned} \quad (5.26)$$

The realization $S(A^b_{n\times n}, b^b_{n\times 1}, c^b_{1\times n})$ is called a balanced realization if its controllability and observability Gramians are equal and diagonal. The diagonal elements of the Gramians in balanced realization will be the Hankel Singular Values. Also, Fernando and Nicholson (1983) proved that the square of the cross-Gramian matrix of the balanced realization is diagonal, and is equal to the product of the controllability and observability Gramians

$$W^2_{cob} = W_{cb}W_{ob} = diag([\sigma_H^{(1)^2} \quad \sigma_H^{(2)^2} \quad \cdots \quad \sigma_H^{(n)^2}]) \quad (5.27)$$

where W_{cob}, W_{cb} and W_{ob} are the cross-Gramian and Gramian Matrices corresponding to the balanced realization.

5.3 Singular Perturbation Based Input-Output Pairing

Singular perturbation is a well established technique to analyze the plants with significantly different time constants. In complex processes with many different control loops, some loops may be faster or slower than the other loops. For example, in typical process control systems, flow, level and pressure control loops are substantially faster than the composition control loops. In such cases, singular perturbation can be employed for plant analysis. The materials in this section are based on the results of (Shimizu and Matsubara 1985).

5.3.1 *Singular Perturbation*

A linear time-invariant plant with significantly different time scales can be singularly perturbed as

$$\dot{x} = A_{11}x + A_{12}p + B_1u \quad (5.28)$$
$$\varepsilon\dot{p} = A_{21}x + A_{22}p + B_2u \quad (5.29)$$
$$y = C_1x + C_2p \quad (5.30)$$

where x and p are $n\times 1$ and $l\times 1$ state vectors, u and y are $m\times 1$ input and output vectors, and ε is a small positive parameter. The matrices in equations (5.28)-(5.30) are constant with appropriate dimensions. Assume that the plant has n slow and l fast eigenvalues of order $O(1)$ and $O(1/\varepsilon)$. Hence, the plant is decomposed into fast and slow subsystems.

Consider the asymptotic expansion for x as

$$x \approx \sum_{k=0}^{\infty} \varepsilon^k x_k = x_0 + \varepsilon x_1 + \cdots \tag{5.31}$$

Let A_{22} be nonsingular, using equations (5.31) and (5.29), and considering the terms of order $O(1)$, we have

$$p = -A_{22}^{-1} A_{21} x_0 - A_{22}^{-1} B_2 u + O(\varepsilon) \tag{5.32}$$

Hence, equation (5.28) can be rewritten as

$$\dot{x}_0 = (A_{11} - A_{12} A_{22}^{-1} A_{21}) x_0 + (B_1 - A_{12} A_{22}^{-1} B_2) u + O(\varepsilon) = A_0 x_0 + B_0 u + O(\varepsilon) \tag{5.33}$$

which is the slow subsystem, and the fast subsystem is

$$\varepsilon \dot{p} = A_{21} x_0 + A_{22} p + B_2 u + O(\varepsilon) \tag{5.34}$$

Combining equations (5.33) and (5.34), gives

$$\begin{bmatrix} \dot{x}_0 \\ \dot{p} \end{bmatrix} = \begin{bmatrix} A_0 & 0 \\ \dfrac{A_{21}}{\varepsilon} & \dfrac{A_{22}}{\varepsilon} \end{bmatrix} \begin{bmatrix} x_0 \\ p \end{bmatrix} + \begin{bmatrix} B_0 \\ \dfrac{B_2}{\varepsilon} \end{bmatrix} u + \begin{bmatrix} O(\varepsilon) \\ O(1) \end{bmatrix} \tag{5.35}$$

and the diagonal form of (5.35) is

$$\begin{bmatrix} \dot{X}_0 \\ \dot{P} \end{bmatrix} = \begin{bmatrix} J_0 & 0 \\ 0 & \dfrac{J_{22}}{\varepsilon} \end{bmatrix} \begin{bmatrix} X_0 \\ P \end{bmatrix} + \begin{bmatrix} U_0^{-1} B_0 \\ U_{22}^{-1} \dfrac{B_2}{\varepsilon} \end{bmatrix} u + \begin{bmatrix} O(\varepsilon) \\ O(1) \end{bmatrix} \tag{5.36}$$

where J_0 and J_{22} are the Jordan form of A_0 and A_{22} with the corresponding U_0 and U_{22} modal matrices, respectively. So, using the new state variables equation (5.30) can be rewritten as

$$y = [C_0 U_0 \quad C_2 U_{22}] \begin{bmatrix} X_0 \\ P \end{bmatrix} + O(\varepsilon) \tag{5.37}$$

where $C_0 = C_1 - C_2 A_{22}^{-1} A_{21}$ and the corresponding transfer function matrix is

$$G(s)=C_0U_0\left(sI-U_0^{-1}A_0U_0\right)^{-1}U_0^{-1}B_0+C_2U_{22}\left(sI-\frac{U_{22}^{-1}A_{22}U_{22}}{\varepsilon}\right)^{-1}U_{22}^{-1}\frac{B_2}{\varepsilon} \tag{5.38}$$

this can be rewritten as

$$G(s)=\Phi(s)+\Phi^{\varepsilon}(s) \tag{5.39}$$

where $\Phi(s)=C_0\left(sI-A_0\right)^{-1}B_0$ and $\Phi^{\varepsilon}(s)=C_2\left(\varepsilon sI-A_{22}\right)^{-1}B_2$.

5.3.2 The Dynamic RGA

According to the RGA definition, relative gain is the ratio of the open loop sensitivity to the closed loop sensitivity as

$$\gamma_{ij}=\left(\frac{\partial y_i}{\partial u_j}\Big|_{u_k=0,k\neq j}\right)\Big/\left(\frac{\partial y_i}{\partial u_j}\Big|_{y_k=0,k\neq i}\right) \tag{5.40}$$

and the RGA is

$$\Gamma=\left[\gamma_{ij}\right]\quad i,j=1,\cdots,m \tag{5.41}$$

Now, let $\bar{y}$ be the steady state output corresponding to the steady state input $\bar{u}$. So, the steady state relation of input-output is

$$\bar{y}=G\bar{u} \tag{5.42}$$

where

$$G=G(0)=-\begin{bmatrix}C_1 & C_2\end{bmatrix}\begin{bmatrix}A_{11} & A_{12}\\ A_{21} & A_{22}\end{bmatrix}^{-1}\begin{bmatrix}B_1\\ B_2\end{bmatrix}$$

Using (5.39) an approximation of system step response for $u(s)=\bar{u}/s$ is

$$y(s)=\left(\Phi(s)+\Phi^{\varepsilon}(s)\right)\frac{\bar{u}}{s} \tag{5.43}$$

and it can be rewritten as

$$y(s) = \left(\Phi(s) + \Phi^{\varepsilon}(s)\right) G^{-1} \frac{\bar{y}}{s} \tag{5.44}$$

Let, λ_k and ν_k be the eigenvalues of A_0 and A_{22} respectively, then we can rewrite equation (5.44) as

$$y(s) = \left\{ \sum_{k=1}^{n} \frac{Q_k}{s - \lambda_k} + \sum_{k=1}^{l} \frac{\frac{R_k}{\varepsilon}}{s - \frac{\nu_k}{\varepsilon}} \right\} G^{-1} \frac{\bar{y}}{s} \tag{5.45}$$

where no multiple eigenvalues are assumed and Q_k and R_k are the corresponding residuals for λ_k and ν_k. Also, these residuals can be computed using equation (5.38), where Q_k is the outer-product of the kth column of $C_0 U_0$ and the kth row of $U_0^{-1} B_0$, and R_k is the outer-product of the kth column of $C_2 U_{22}$ and the kth row of $U_{22}^{-1} B_2$.

Let the ith element of $y(s)$ and $\bar{y}$ be $y_i(s)$ and $\bar{y}_i$, respectively. Hence, equation (5.45) gives

$$\frac{y_i(s)}{\bar{y}_i} = \sum_{j=1}^{m} \tilde{\gamma}_{ij}(s) \tag{5.46}$$

where

$$\tilde{\gamma}_{ij}(s) - \sum_{k=1}^{n} \frac{q_{kij}\, \hat{g}_{ji}}{s(s - \lambda_k)} + \sum_{k=1}^{l} \frac{r_{kij}\, \hat{g}_{ji}}{s(\varepsilon s - \nu_k)} \tag{5.47}$$

It is obvious that q_{kij}, r_{kij} and $\hat{g}_{ij}$ are the $(i,j)^{th}$ component of Q_k, R_k and G^{-1}, respectively. Note that according to the relative gain definition in equation (5.40), $\tilde{\gamma}_{ij}(s)$ is the relative gain corresponding to u_j and y_i.

Also, equation (5.45) can be computed in the time domain as

$$\frac{y_i(t)}{\bar{y}_i} = \sum_{j-1}^{m} \tilde{\gamma}_{ij}(t) \tag{5.48}$$

where

$$\tilde{\gamma}_{ij}(t) = \sum_{k=1}^{n} \frac{q_{kij}\, g_{ji}}{\lambda_k} (e^{\lambda_k t} - 1) + \sum_{k=1}^{l} \frac{r_{kij}\, g_{ji}}{\nu_k} (e^{\left(\frac{\nu_k}{\varepsilon}\right) t} - 1) \tag{5.49}$$

Note. In fact, for computational purposes the relative gains, $\tilde{\gamma}_{ij}(t)$ can be computed using the step responses of $\left(\Phi(s)+\Phi^{\varepsilon}(s)\right).*G^{-T}$.

The Singular Perturbation based pairing rules. As stated earlier, singular perturbation is an appropriate input-output pairing approach for plants with significantly different time scales. Along with the RGA pairing rules introduced in chapter two, the following points are to be considered:

- Choose the input-output pair with $\tilde{\gamma}_{ij}(t)$ closer to 1.
- Relative gains $\tilde{\gamma}_{ij}(t)$ corresponding to the appropriate input-output pair should be positive.

Example 5.3.1

Consider a three-stage distillation column with the following state space model

$$\dot{x} = \begin{bmatrix} -14 & 12 & 0 \\ 23 & -51 & 35 \\ 0 & 5 & -11 \end{bmatrix} x + \begin{bmatrix} 0 \\ -10 \\ -9 \end{bmatrix} p + \begin{bmatrix} 0 & 0 & 0 \\ 3.8 & -2.9 & 0 \\ 0.59 & -2.9 & 0 \end{bmatrix} u$$

$$\varepsilon \dot{p} = \begin{bmatrix} 0 & 10 & 3 \end{bmatrix} x + \begin{bmatrix} -4 \end{bmatrix} p + \begin{bmatrix} -0.2 & -2.2 & 3.3 \end{bmatrix} u$$

$$y = \begin{bmatrix} 1 & 0 & 0 \\ 0 & 0 & 1 \\ 0 & 0 & 0 \end{bmatrix} x + \begin{bmatrix} 0 \\ 0 \\ 1 \end{bmatrix} p$$

and the corresponding RGA is

$$\Lambda = \begin{bmatrix} 0.8012 & 0.2787 & -0.0799 \\ 0.1618 & -0.3167 & 1.1549 \\ 0.0370 & 1.0380 & -0.0750 \end{bmatrix}$$

where it proposes the $(u_1 - y_1, u_2 - y_3, u_3 - y_2)$ as the appropriate input-output pair. To use the singular perturbation pairing method all the nine relative gains $\tilde{\gamma}_{ij}(t),\ i,j = 1,2,3$ must be calculated. To show the main ideas in this example, only the relative gain between the 3rd output and all the inputs is considered. Figures 5.1 and 5.2 show the time behavior of $\tilde{\gamma}_{3j}(t),\ j = 1,2,3$ for $\varepsilon = 5$ and $\varepsilon = 0.5$. Figures 5.1 and 5.2 show that for $\varepsilon = 5$, the second input significantly affects the 3rd output in the steady state, i.e. $\tilde{\gamma}_{32}(\infty) >> \tilde{\gamma}_{33}(\infty)$, while in the transient response the 3rd input has significant effect on the 3rd output. Note that $\tilde{\gamma}_{31}(t)$ is small for all t. So, the plant is highly interactive. But, for $\varepsilon = 0.5$ we have for most of the times $\tilde{\gamma}_{32}(t) >> \tilde{\gamma}_{33}(t)$, and the plant dynamic interaction is reduced. In this case, $(u_2 - y_3)$ is the appropriate input-output pair.

■

5.4 The Participation Matrix

In this section, the controllability and observability Gramian matrices of stable multivariable plants are used to select input-output pairs based on the plant state space model.

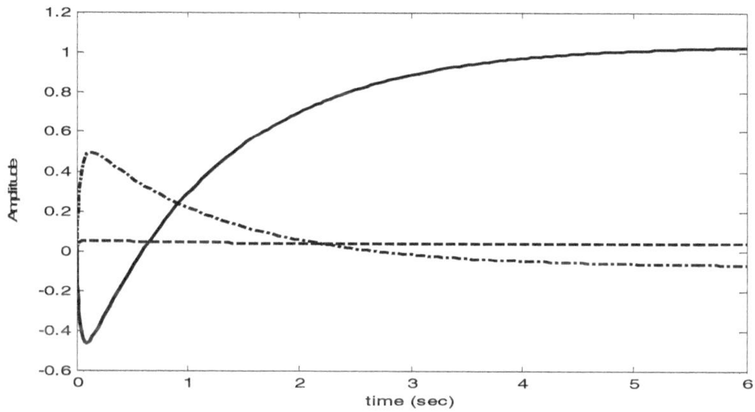

Fig. 5.1 Time courses of $\tilde{\gamma}_{3j}(t)$, $j = 1,2,3$ for $\varepsilon = 5$. Solid line corresponds to $\tilde{\gamma}_{32}(t)$, Dash-dot line correspond to $\tilde{\gamma}_{33}(t)$ and dashed line corresponds to $\tilde{\gamma}_{31}(t)$.

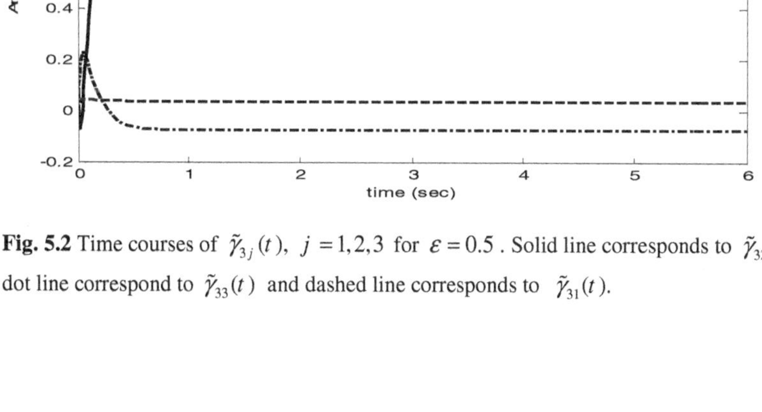

Fig. 5.2 Time courses of $\tilde{\gamma}_{3j}(t)$, $j = 1,2,3$ for $\varepsilon = 0.5$. Solid line corresponds to $\tilde{\gamma}_{32}(t)$, Dash-dot line correspond to $\tilde{\gamma}_{33}(t)$ and dashed line corresponds to $\tilde{\gamma}_{31}(t)$.

5.4.1 Gramians in Multivariable Plants

Consider the $m \times m$ transfer function matrix $G(s)$, with corresponding state space description denoted by $S(A,B,C,0)$, where $B = [b_1 \;\; b_2 \;\; \ldots \;\; b_m]$ and $C^T = [c_1^T \;\; c_2^T \;\; \ldots \;\; c_m^T]$. Then, the elementary subsystems can be described as $S(A,b_i,c_j,0)$, which are the SISO plants that build the original MIMO plant, $G(s) = [G_{ij}(s)]$, $i,j = 1,2,\ldots,m$. According to the Gramians definition in 5.2.1, each stable elementary plant has the controllability and observability Gramians satisfying the following equations

$$W_{c,j}A^T + AW_{c,j} = -b_j b_j^T \tag{5.50}$$

and

$$W_{o,i}A + A^T W_{o,i} = -c_i^T c_i \tag{5.51}$$

Moreover, the Gramian matrices can be computed for the original MIMO plant by solving the following Lyapunov equations

$$W_c A^T + AW_c = -BB^T \tag{5.52}$$

and

$$W_o A + A^T W_o = -C^T C \tag{5.53}$$

In (Salgado and Conley 2004) it is shown that the controllability and observability Gramians of the original MIMO plant are equal to the summation of the corresponding Gramians of the elementary subsystems. That is

$$W_c = \sum_{j=1}^{m} W_{c,j} \tag{5.54}$$

and

$$W_o = \sum_{i=1}^{m} W_{o,i} \tag{5.55}$$

It is straightforward to prove that Gramian matrices of the elementary subsystems and the corresponding original MIMO plant satisfy the following equation (Salgado and Conley 2004)

$$W_c W_o = \sum_{i=1}^{m} \sum_{j=1}^{m} W_{c,j} W_{o,i} \tag{5.56}$$

Equation (5.56) expresses the fact that all SISO subsystems have an effect on the controllability and observability of the original plant.

5.4.2 The Interaction Measure

Studying equation (5.56) reveals that the matrix products $W_{c,j}W_{o,i}$, show the effect of the elementary subsystems corresponding to the jth input and the ith output in the controllability and observability of overall multivariable plant. As mentioned in section 5.2.1, the eigenvalues of this product are independent of the state space realization and all its eigenvalues are nonnegative. Hence, the eigenvalues sum can be an appropriate index of the controllability and observability degrees corresponding to the elementary subsystems and it can be used as a measure of interaction. It is intuitively clear that this summation is equal to the trace of the matrix $W_{c,j}W_{o,i}$, and is denoted by $trace(W_{c,j}W_{o,i})$. We have

$$trace(W_{c,j}W_{o,i}) = \sum_{k=1}^{n} \lambda_k = \sum_{k=1}^{n} \sigma_H^{(k)^2} \tag{5.57}$$

where λ_k and $\sigma_H^{(k)}$ denote the kth eigenvalue of $W_{c,j}W_{o,i}$ and the kth Hankel singular value, respectively. Thus, equation (5.57) shows that HSVs of an elementary subsystem G_{ij}, quantify the combined abilities of the input u_j and the output y_i to control and observe the plant states.

Salgado and Conley (2000) and (2004) have defined the *Participation Matrix* (PM) as

$$\Phi = \left[\varphi_{ij} \right] \tag{5.58}$$

where

$$\varphi_{ij} = \frac{trace(W_{c,j}W_{o,i})}{trace(W_c W_o)} \tag{5.59}$$

In equation (5.59), $trace(W_c W_o)$ is used to normalize the interaction measure. It results in $0 < \varphi_{ij} < 1$ and

$$\sum_{i=1}^{m} \sum_{j=1}^{m} \varphi_{ij} = 1 \tag{5.60}$$

5.4.3 The Participation Matrix Properties

The PM has several important properties pertinent to input-output pairing analysis. Some of these properties are summarized below.

Property 1. The participation matrix is sensitive to input and output scaling. For example, if $G_{ij}(s)$ is multiplied by a gain k, then $trace(W_{c,j}W_{o,i})$ will be multiplied by k^2. One way to deal with this issue is discussed in (Glad and Ljung 2000) and (Maciejowski 1989).

Property 2. The PM is independent of frequency scaling. In the other word, $trace(W_{c,j}W_{o,i})$ for $G_{ij}(s)$ and $G_{ij}(s/\varepsilon)$ are equal, where $\varepsilon \in \Re^+$. This implies that the $trace(W_{c,j}W_{o,i})$ can be computed for standard plants, as shown in table 5.1, and applied to a wide range of plants. On the other hand, this property can be regarded as a deficiency of the PM interaction measure, since it cannot distinguish the fast and slow poles in a similar class of transfer functions.

Table 5.1 shows the following points concerning $trace(W_{c,j}W_{o,i})$ computation, summarized as:

- For the first order plants, $trace(W_{c,j}W_{o,i})$ is determined by its steady state gain.
- For the second order plants without zero, two distinct cases can be considered.
  - For $0<\alpha<2$, $trace(W_{c,j}W_{o,i})$ grows as α decreases. So, inputs and outputs corresponding to oscillatory modes are strong pairing candidates.
  - For $2<\alpha$, the second order plant can be separated into two first order terms.
- For the second order plants with a zero, if $\beta \to 0$ we get similar results to the plants with no zeros. If $\beta \to \infty$ or $\beta < 0$, then $trace(W_{c,j}W_{o,i})$ grows, but this does not imply a pairing selection.

Table 5.1 $trace(W_{c,j}W_{o,i})$ for standard first and second order plants.

Transfer Function, $G_{ij}(s)$	$trace(W_{c,j}W_{o,i})$
$k\dfrac{1}{s+1}$	$k^2\dfrac{1}{4}$
$k\dfrac{1}{s^2+\alpha s+1}$	$k^2\dfrac{2+\alpha^2}{4\alpha^2}$
$k\dfrac{\beta s+1}{s^2+\alpha s+1}$	$k^2\dfrac{2+\alpha^2+2\beta^2-2\alpha\beta}{4\alpha^2}$

Note. In the cases where a fully decentralized controller results in unacceptably poor closed loop performance, the PM can be used to propose a richer sparse or block diagonal controller structure.

Although the plants in table 5.1 are simple, they model a vast number of multivariable plants. However, industrial processes are commonly time delayed. In (Salgado and Conley 2004) it is shown that by transforming the continuous model to their equivalent discrete-time form, and computing the corresponding Gramians the effect of time delay can be considered.

Consider the time delayed SISO stable system S_l, separated as $S_o(A_o, b_o, c_o, 0)$, and a pure delay l, where $l \in \mathbf{N}$ as shown in figure 5.3. The Gramian matrices of the discrete-time state space model S_o satisfy the following equations

$$W_c^o - A_o W_c^o A_o^T - b_o b_o^T = 0 \tag{5.61}$$

and

$$W_o^o - A_o^T W_o^o A_o - c_o^T c_o = 0 \tag{5.62}$$

and if W_c^l and W_o^l represent the controllability and observability Gramians of S_l, the $trace(W_c^l W_o^l)$ can be computed as (Salgado and Conley 2004)

$$trace(W_c^l W_o^l) = trace(W_c^o W_o^o) + l c_o W_c^o c_o \tag{5.63}$$

where using equation (5.7), this equation can be rewritten as

$$trace(W_c^l W_o^l) = trace(W_c^o W_o^o) + l \left\| G_o \right\|_2^2 \tag{5.64}$$

Using equation (5.64), the PM can be evaluated for time delayed multivariable plants.

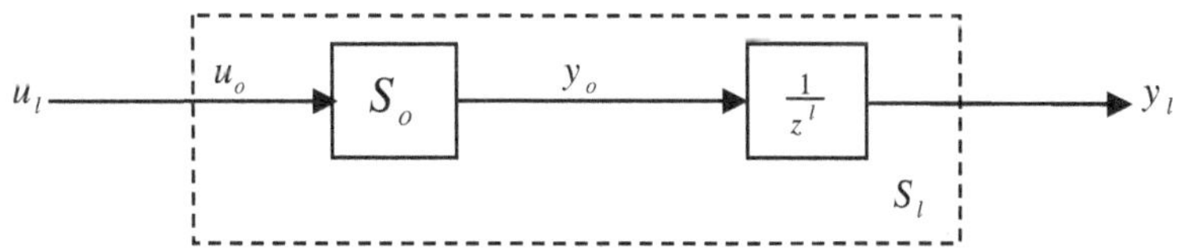

Fig. 5.3 Discrete-time plant with time delay

The Participation Matrix based pairing rules. These rules are summarized as follows:

- Select the input-output pair with a large φ_{ij} or φ_{ij}s greater than the average ($\varphi_{ij} \gg 1/m^2$).
- Select the input-output pair such that the sum of the corresponding φ_{ij} is close to one. In highly interactive plants choose the block diagonal structure.

Example 5.4.1

Consider a 3×3 multivariable plant with the PM given by

$$\Phi = \begin{bmatrix} 0.1697 & 0.0782 & 0.0282 \\ 0.1979 & 0.1800 & 0.0801 \\ 0.0896 & 0.0716 & 0.1046 \end{bmatrix}$$

the average value of the PM elements is $1/9 = 0.1111$. The PM pairing rules propose the pairs with φ_{ij} greater than 0.1111. Only three elements of the PM, φ_{11}, φ_{21} and φ_{22}, satisfy this condition.

Note that φ_{21} corresponds to the largest element of Φ, and $(u_1 - y_2)$ is selected as the appropriate pair. The other input-output pairs will be $(u_2 - y_1, u_3 - y_3)$ and the sum of the corresponding PM elements is 0.3807. But, $(u_1 - y_1, u_2 - y_2, u_3 - y_3)$ can be alternatively selected as the appropriate pair and the corresponding PM elements sum is 0.4543, which would be a more desirable control configuration for this plant.

■

Example 5.4.2

Consider the following 3×3 multivariable plant (Salgado and Conley 2004)

$$G(s) = \begin{bmatrix} \dfrac{0.4}{(s+1)^2} & \dfrac{4(s+3)}{(s+2)(s+5)} & \dfrac{-2}{s+4} \\ \dfrac{2}{(s+2)(s+1)} & \dfrac{2}{(s+2)^2} & \dfrac{1}{s+2} \\ \dfrac{6(-s+1)}{(s+5)(s+4)} & \dfrac{4}{(s+3)^2} & \dfrac{8}{(s+2)(s+5)} \end{bmatrix}$$

Its corresponding PM is

$$\Phi = \begin{bmatrix} 0.0370 & 0.2018 & 0.0385 \\ 0.2226 & 0.0578 & 0.0385 \\ 0.2193 & 0.0457 & 0.1389 \end{bmatrix}$$

It is obvious that φ_{12}, φ_{21} and φ_{33} are the best PM elements for pairing with the maximum summation. Hence, $(u_1 - y_2, u_2 - y_1, u_3 - y_3)$ is the proposed control configuration selection.

■

Example 5.4.3

Consider the 2×2 discrete-time transfer function matrix (Salgado and Conley 2004)

$$G(z) = \begin{bmatrix} \dfrac{0.5}{z-0.5} & \dfrac{0.15}{(z-0.8)z^{l}} \\ \dfrac{0.1}{(z-0.5)(z-0.8)} & \dfrac{0.3}{z-0.7} \end{bmatrix}$$

where l is a nonnegative integer and indicates the time delay in the G_{12} element. Here, the PM is computed in two different cases $l = 0$ and $l = 10$. The corresponding participation matrices are

$$\Phi|_{l=0} = \begin{bmatrix} 0.3171 & 0.1239 \\ 0.3122 & 0.2469 \end{bmatrix}$$

and

$$\Phi|_{l=10} = \begin{bmatrix} 0.2193 & 0.3941 \\ 0.2195 & 0.1707 \end{bmatrix}$$

Hence, the appropriate input-output pairs are selected as $(u_1 - y_1, u_2 - y_2)$ and $(u_1 - y_2, u_2 - y_1)$, respectively. This clearly shows that the time delay change from zero to 10 steps will change the proposed control configuration.

■

5.5 The Hankel Interaction Index Arra

Singular perturbation and the Participation Matrix pairing methods consider the plant dynamics via its state space model. In the PM, sum of the squared Hankel singular values for the elementary subsystems is used to measure the interaction.

In this section, the Hankel interaction index array (HIIA) is introduced that uses the Hankel norm of the elementary subsystems to quantify the plant interaction. This provides physical insight to the pairing process and uses the full frequency range of the plant. The presented materials are based on the results of (Wittemark and Salgado 2002).

5.5.1 The Hankel Interaction Measure

In section 5.2, the Gramian matrices and HSV were introduced. Also, the Hankel norm of a plant was defined as the maximum HSV. Hankel norm is the 2-induced norm of Hankel operator which assigns past inputs u_- to future outputs y_+ as

$$y_+ = \mathcal{H}(u_-)(t) = \int_{-\infty}^{0} h(t-\tau)u_-(t)d\tau$$

where $\mathcal{H}$ denotes the Hankel operator (Antoulas 2005). For stable multivariable plants with the transfer function matrix $G(s)$, the Hankel norm is defined as (Skogestad and Postlethwaite 2005)

$$\|G(s)\|_H \triangleq \sup_{u\neq 0} \frac{\sqrt{\int_0^{\infty} \|y(\tau)\|_2^2 d\tau}}{\sqrt{\int_{-\infty}^{0} \|u(\tau)\|_2^2 d\tau}} \tag{5.65}$$

Equation (5.65) shows the effect of an input on the states and how much that effect is reflected in the output. On the other hand, the Hankel norm indicates the effect of the past inputs on future outputs which can be useful for input-output pairing.

To define a new interaction index, for each elementary subsystem $S(A, b_j, c_i, 0)$ the Hankel norm is computed to measure the ability of input u_j to control the output y_i. So, a matrix which contains the Hankel norm of subsystems is introduced to indicate the plants inputs and outputs relations and interactions. Let

$$\bar{\Sigma}_H = \left[\|G_{ij}(z)\|_H\right] \tag{5.66}$$

which is scaling dependant, and is further modified to get the scaling independent Hankel Interaction Index Array (HIIA) defined as follows

$$\Phi = [\varphi_{ij}] = \left[\frac{\left\| G_{ij}(z) \right\|_H}{\sum_{i,j} \left\| G_{ij}(z) \right\|_H} \right] \tag{5.67}$$

The HIIA based pairing rules. Employing the HIIA, the following input-output pairing rule is proposed:

- Choose the input-output pair corresponding to the largest element in each row of the HIIA, and then remove the corresponding row and column. This rule should be repeated until all rows and columns are removed or pairing is finished.

Example 5.5.1

Consider the Wood and Berry (1973) distillation column presented in example 2.2.1

$$G(s) = \begin{bmatrix} \dfrac{12.8e^{-s}}{16.7s+1} & \dfrac{-18.9e^{-s}}{21s+1} \\ \dfrac{6.6e^{-7s}}{10.9s+1} & \dfrac{-19.4e^{-s}}{14.4s+1} \end{bmatrix}$$

Using a 1st order Pade approximation for the delays, we have

$$\bar{\Sigma}_H = \begin{bmatrix} 6.7527 & 9.8708 \\ 4.4813 & 12.6164 \end{bmatrix}$$

also, the corresponding HIIA using equation (5.67) is

$$\Phi = \begin{bmatrix} 0.2002 & 0.2927 \\ 0.1329 & 0.3741 \end{bmatrix}$$

It is obvious that the HIIA proposes $(u_1 - y_1, u_2 - y_2)$ as the appropriate input-output pair. This is analogous to the result of the RGA analysis in example 2.2.1.

■

Example 5.5.2

Consider the following transfer function matrix (Wittemark and Salgado 2002)

$$G(z)=\begin{bmatrix} \dfrac{0.1021}{z-0.9048} & \dfrac{0.3707z-0.3535}{z^2-1.724z+0.7408} \\ \dfrac{-0.192z+0.1826}{z^2-1.869z+0.8781} & \dfrac{0.09516}{z-0.9048} \end{bmatrix}$$

The corresponding RGA is

$$\Lambda=\begin{bmatrix} 0.5034 & 0.4966 \\ 0.4966 & 0.5034 \end{bmatrix}$$

and it proposes the (u_1-y_1, u_2-y_2) as the appropriate input-output pair. But, the HIIA is

$$\Phi=\begin{bmatrix} 0.1936 & 0.2978 \\ 0.3281 & 0.1805 \end{bmatrix}$$

and the off-diagonal pairing is recommended. To further study this contradictory pairing results, a dead-beat controller which drives the error to zero is designed in (Wittenmark and Salgado 2002) as follows

$$C(z)=(I-G_o(z)Q(z))^{-1}Q(z) \tag{5.68}$$

where $Q(z)$ is any stable transfer function and $G_o(z)$ is the nominal plant which the appropriate input-output pair assigns its structure. That is, for diagonal pairing we have

$$G_o(z)=\begin{bmatrix} G_{11}(z) & 0 \\ 0 & G_{22}(z) \end{bmatrix}$$

and for off-diagonal pairing

$$G_o(z)=\begin{bmatrix} 0 & G_{12}(z) \\ G_{21}(z) & 0 \end{bmatrix}$$

To achieve the desired performance, $Q(z)$ is chosen as

$$Q(z)=\frac{1}{z^2}(G_o(z))^{-1}$$

Using this controller for the diagonal pairing, the closed loop plant will be unstable, while the closed loop responses corresponding to the off-diagonal pairing are shown in figure 5.4. This reveals that $(u_1 - y_2, u_2 - y_1)$ is the appropriate input-output pair.

■

5.6 The Dynamic Input-Output Pairing Matrix

Two main characteristics of an effective input-output pairing method are simplicity and ability to encompass the dynamical behaviour of the plant. Both the PM and HIIA methods analyze the dynamical interaction, and employ the results in their pairing selection. Interaction is evaluated using the controllability and observability Gramian matrices for each elementary subsystem. This makes their application rather complicated. In this section, a less computationally demanding approach based on the cross-Gramian matrix is introduced based on the results in (Moaveni and Khaki Sedigh 2008).

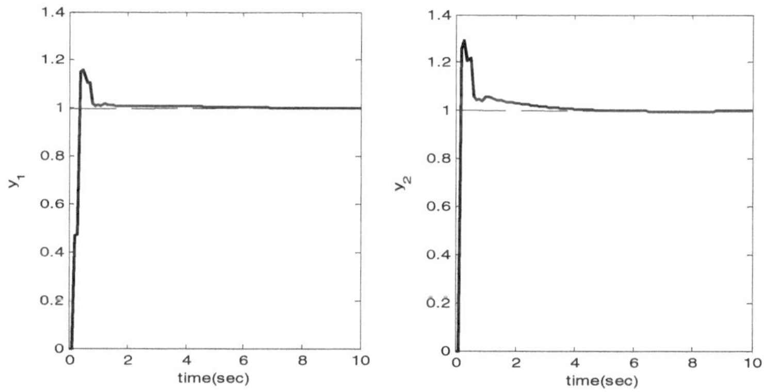

Fig. 5.4 Closed loop step responses under decentralized control in example 5.5.2, corresponding to off-diagonal pairing proposed by the HIIA.

5.6.1 Input-Output Pairing Using the Cross-Gramian Matrix

Let (A^b, B^b, C^b) be a balanced realization of the linear stable $m \times m$ transfer function matrix $G(s)$. Also, (A^b, b_j^b, c_i^b) are considered as the elementary subsystems with balanced realization defined for $i, j = 1, 2, ..., m$. Each elementary subsystem has a corresponding diagonal cross-Gramian matrix W_{cob}^{ij} and its norm defined as the largest singular value $\bar{\sigma}(W_{cob}^{ij})$ can be employed to quantify the ability of input u_j to control the output y_i. It is obvious that the largest singular value of W_{cob}^{ij} is the Hankel norm of (A^b, b_j^b, c_i^b), where the Hankel norm properties and its relation with the plant inputs-output were introduced in the previous sections. The Dynamic Input-Output Pairing Matrix (DIOPM) is defined as

$$\Phi = \left[\varphi_{ij}\right] = \left[\bar{\sigma}(W_{cob}^{ij})\right] \tag{5.69}$$

Equation (5.69) states that the DIOPM is calculated using the balanced realization of the plant. However, it is shown in (Moaveni and Khaki Sedigh 2008) that $\bar{\sigma}(W_{cob}^{ij})$ is equal to the maximum of the absolute values of the eigenvalues of the cross-Gramian for any realization. In fact, the DIOPM can be rewritten as

$$\Phi = \left[\bar{\sigma}(W_{cob}^{ij})\right] = \left[\sqrt{\lambda_{\max}((W_{cob}^{ij})^2)}\right] = \left[\max\{|\lambda(W_{co}^{ij})|\}\right] \tag{5.70}$$

where λ denotes the matrix eigenvalues. This substantially reduces the computational load in input-output pairing for large scale plants and provides physical insight to the input-output relationships as φ_{ij} is the Hankel norm of subsystems.

The DIOPM based pairing rules. Employing the DIOPM, the following pairing rule is proposed:

- Use the cross-Gramian matrices corresponding to the elementary subsystems of the plant to compute the corresponding Hankel norm. Then form the DIOPM and select the largest elements of Φ to identify the control configuration.

Note. A similar method is introduced in (Khaki-Sedigh and Shahmansourian 1996) that uses the determinant of the W_{cob}^{ij} for pairing analysis. However, it does not provide any physical insight as in the Hankel norm.

Example 5.6.1

Consider the plant with transfer function matrix

$$G(s) = \begin{bmatrix} \dfrac{-0.9019s + 15.47}{s^2 + 9.163s + 15.47} & \dfrac{-3.327}{s + 6.931} \\ \dfrac{0.8926}{s + 2.231} & \dfrac{0.7549s + 13.92}{s^2 + 9.163s + 15.47} \end{bmatrix}$$

Corresponding RGA for this system is

$$\Lambda = \begin{bmatrix} 0.8241 & 0.1759 \\ 0.1759 & 0.8241 \end{bmatrix}$$

and the DIOPM is

$$\Phi = \begin{bmatrix} 0.6182 & 0.2400 \\ 0.2000 & 0.4915 \end{bmatrix}$$

It is easily seen that both RGA and DIOPM propose $(u_1 - y_1, u_2 - y_2)$, as the appropriate input-output pair.

■

Example 5.6.2

Consider the process given in (Grosdidier and Morari 1986) as

$$G(s) = \begin{bmatrix} \dfrac{5}{4s+1} & \dfrac{2.5e^{-5s}}{(2s+1)(15s+1)} \\ \dfrac{-4e^{-6s}}{20s+1} & \dfrac{1}{3s+1} \end{bmatrix}$$

The corresponding RGA is

$$\Lambda = \begin{bmatrix} 0.333 & 0.667 \\ 0.667 & 0.333 \end{bmatrix}$$

which proposes the $(u_1 - y_2, u_2 - y_1)$ as the appropriate input-output pair. But, the DIOPM using (5.70) is

$$\Phi = \begin{bmatrix} 2.5000 & 1.6049 \\ 2.4336 & 0.5000 \end{bmatrix}$$

where the 3rd order Pade approximation is used to realize the delay. The DIOPM shows that $(u_1 - y_1, u_2 - y_2)$ is the appropriate input-output pair. This loop pairing decision was reported in (Grosdidier and Morari 1986), through analyzing both magnitude and phase characteristics of the interaction between the two loops and by Xiong and Cai in (Xiong et al. 2005) using the ERGA.

■

5.7 Conclusion

The main purpose of this chapter has been to provide an overview of the control configuration selection methodologies based on the state space model of linear multivariable plants.

The controllability and observability Gramians definitions and computations were presented in section 5.2. Then, a new method to compute the cross-Gramian matrix by solving the Sylvester equation is introduced. In section 5.3, singular perturbation approach is introduced that can handle multivariable plants with significantly different time constants. In subsequent sections Participation Matrix (PM), Hankel Interaction Index Array (HIIA) and Dynamic Input-Output Pairing Matrix (DIOPM) based on the Hankel Singular Values (HSV) of the elementary subsystems are introduced. Examples are provided to show the main points of the proposed pairing methodologies.

References

Antoulas, A.C.: Approximation of large-scale dynamical systems. SIAM publications, Philadelphia (2005)

Doyle, J., Francis, B., Tannenbaum, A.: Feedback control Theory. Macmillan Publishing Co., New York (1992)

Fernando, K.V., Nicholson, H.: On the structure of balanced and other principal representations of SISO systems. IEEE Trans. Autom. Control 28, 228–231 (1983)

Glad, T., Ljung, L.: Control theory, multivariable and nonlinear methods. Taylor and Francis, London (2000)

Grosdidier, P., Morari, M.: Interaction measures for systems under decentralized control. Automatica 22, 309–319 (1986)

Khaki-Sedigh, A., Shahmansourian, A.: Input-output pairing using balanced realizations. Electron. Lett. 32, 2027–2028 (1996)

Maciejowski, J.: Multivariable feedback design. Addison Wesley, Wokingham (1989)

Moaveni, B., Khaki-Sedigh, A.: A new approach to compute the cross-Gramian matrix and its application in input-output pairing of linear multivariable plants. J. Applied Sciences 8, 608–614 (2008)

Salgado, M.E., Conley, A.: MIMO interaction measure and controller structure selection. Int. J. Control 77, 367–383 (2004)

Shimizu, K., Matsubara, M.: Singular perturbation for the dynamic interaction measure. IEEE Trans. Autom. Control 30, 790–792 (1985)

Skogestad, S., Postlethwaite, I.: Multivariable feedback control analysis and design. Wiley, Chichester (2005)

Wittenmark, B., Salgado, M.E.: Hankel-norm based interaction measure for input-output pairing. In: Proceeding of the 15th IFAC World Congress on Automatic Control, Barcelona, Spain (2002)

Xiong, Q., Cai, W.J., He, M.J.: A practical loop pairing criterion for multivariable processes. J. Process Contr. 15, 741–747 (2005)

Chapter 6
Control Configuration Selection of Nonlinear Multivariable Plants

6.1 Introduction

In spite of the extensive research of the previous decades in the field of input-output pairing for linear multivariable plants, the input-output pairing problem of nonlinear multivariable plants has received little attention. There are two main approaches to input-output pairing selection of nonlinear multivariable plants. The first and the most widely used approach is to employ the well established linear techniques for the linearized plant model. The second approach, which is the subject of the present chapter, directly uses the nonlinear plant model and derives the appropriate input-output pair either from direct nonlinear input-output relationship or from the nonlinear extensions of the linear concepts by introducing the nonlinear RGA. The materials of this chapter are based on (Daoutidis and Kravaris 1992), (Glad 1999) and (Moaveni and Khaki-Sedigh 2007) and covers the class of affine nonlinear multivariable plants. Worked examples and simulation results are provided to show the effectiveness of the nonlinear pairing methodologies, and limitations of the classical RGA.

Section 6.2 presents some relevant known definitions and results from the nonlinear system theory that is used in the subsequent sections in this chapter. Section 6.3 introduces the digraph of an affine nonlinear multivariable plant and the corresponding structural interaction matrix. This methodology is based on the definition of plant relative order. By presenting a key theorem, the input-output pairing rules for nonlinear plants are derived (Daoutidis and Kravaris 1992). The extended and nonlinear RGA are introduced in section 6.4. The extended RGA is an extension of the definition of the classical RGA based on (Glad 1999). And, the nonlinear RGA is based on the materials from (Moaveni and Khaki-Sedigh 2007). The pairing rules in both cases are similar to the classical RGA. Linear interpretation of the nonlinear RGA is given. Finally, several worked examples are provided to show the effectiveness of the direct nonlinear input-output pairing.

6.2 Elements of the Nonlinear System Theory

In this section, some fundamental results from the nonlinear system theory are briefly reviewed. Nonlinear plants do not posses the linear properties such as, *the*

A. Khaki-Sedigh and B. Moaveni: Control Configuration Selection, LNCIS 391, pp. 139–172.
springerlink.com

response type independence *of the input amplitude and initial conditions* and *the superposition principle.* The main nonlinear behaviors of nonlinear plants are briefly pointed out as follows (Gelb and Vander Velde 1968):

- Nonlinear plants have multiple isolated equilibrium points.
- Nonlinear plant response can oscillate with fixed amplitude and fixed period without any external excitation in plant inputs, where these oscillations are called limit cycles.
- Nonlinear plant output is sensitive to initial conditions.
- A slight change in the input amplitude or frequency of the nonlinear plant can cause the plant to change states.

Nonlinear plants can be classified in terms of their mathematical properties, as *continuous* and *discontinuous*. Discontinuous nonlinearities are also called *hard nonlinearities* (such as a backlash or hystersis). However, the present input-output pairing methods only consider the nonlinear continuous plants.

Nonlinear continuous plants can be described by the following nonlinear differential and algebraic equations

$$\begin{aligned} \dot{x}(t) &= f\left(x(t), u(t)\right) \\ y(t) &= h\left(x(t), u(t)\right) \end{aligned} \tag{6.1}$$

where $x \in \Re^n$, $u \in \Re^m$ and $y \in \Re^m$ denote the state, input and output vectors, respectively. The class of *affine*, *input-linear* or *companion* nonlinear plants is described by the following equations

$$\begin{aligned} \dot{x}(t) &= f\left(x(t)\right) + g\left(x(t)\right)u(t) \\ y(t) &= h\left(x(t)\right) \end{aligned} \tag{6.2}$$

where f and g are smooth vector fields and h is a smooth vector function. The above affine model can be written as

$$\begin{aligned} \dot{x}(t) &= f\left(x(t)\right) + \sum_{j=1}^{m} g_j\left(x(t)\right)u_j(t) \\ y_1(t) &= h_1\left(x(t)\right) \\ &\vdots \\ y_m(t) &= h_m\left(x(t)\right) \end{aligned} \tag{6.3}$$

It is interesting to note that many practical and industrial nonlinear plants are indeed modeled by affine nonlinear equations given by equations (6.3) (Nijmeijer and Van de Schaft 1996) and (Slotine and Li 1991).

We now briefly review some basic definitions and mathematical tools from differential geometry and topology, required in the rest of the chapter.

Definition 6.1. The *vector field* $f:\Re^n \rightarrow \Re^n$ is defined as a vector function that maps an n-dimensional space to another n-dimensional space. Also, a *smooth* vector field is the vector-function $f(x)$ with continuous partial derivatives of any order.

In addition, the *gradient* of a given scalar function $h(x)$, $h:\Re^n \rightarrow \Re$, denoted by ∇h, is a row-vector as

$$\nabla h = \frac{\partial h}{\partial x} = \begin{bmatrix} \frac{\partial h}{\partial x_1} & \cdots & \frac{\partial h}{\partial x_i} & \cdots & \frac{\partial h}{\partial x_n} \end{bmatrix} \tag{6.4}$$

Similarly, the *Jacobian* of a given vector field $f(x)$, $f:\Re^n \rightarrow \Re^n$, is denoted by ∇f and is computed as the following $n \times n$ matrix

$$\nabla f = \frac{\partial f}{\partial x} = \begin{bmatrix} \frac{\partial f_1}{\partial x_1} & \cdots & \frac{\partial f_1}{\partial x_n} \\ & \ddots & \\ \vdots & \frac{\partial f_i}{\partial x_i} & \vdots \\ & \ddots & \\ \frac{\partial f_n}{\partial x_1} & \cdots & \frac{\partial f_n}{\partial x_n} \end{bmatrix}_{n \times n} \tag{6.5}$$

Definition 6.2. For the scalar function $h(x)$, and the vector field $f(x)$, *Lie derivative* of h with respect to f represented by $L_f h$ is a scalar function defined as

$$L_f h = \nabla h . f \tag{6.6}$$

Thus, the Lie derivative $L_f h$ is simply the directional derivative of h along the direction of f. The following points will be useful:

- Repeated Lie derivatives can be defined recursively as

$$\begin{aligned} L_f^0 h &= h \\ L_f^i h &= L_f(L_f^{i-1} h) = \nabla(L_f^{i-1} h).f \end{aligned} \tag{6.7}$$

- Similar to the Lie derivative definition, for a vector field g, a scalar function $L_g L_f h$ can be computed as

$$L_g L_f h == \nabla(L_f h).g \tag{6.8}$$

- Consider the single output nonlinear plant

$$\begin{aligned} \dot{x} &= f(x) \\ y &= h(x) \end{aligned} \tag{6.9}$$

derivatives of the output are

$$\begin{aligned} \dot{y} &= \frac{\partial h}{\partial x}.\dot{x} = \nabla h.f = L_f h \\ \ddot{y} &= \frac{\partial [L_f h]}{\partial x}.\dot{x} = L_f^2 h \end{aligned} \tag{6.10}$$

Definition 6.3. For the nonlinear multivariable plants described by equations (6.3), the *relative order* r_{ij} of the output y_i with respect to the input u_j is defined as the smallest integer for which

$$L_{g_j} L_f^{r_{ij}-1} h_i(x) \neq 0 \tag{6.11}$$

If there is no integer to satisfy the above equation then the corresponding relative order will be infinity, $r_{ij} = \infty$ (Isidori 1995).

Note. It can be shown that for finite r_{ij}, $r_{ij} \leq n$.

According to definition 6.3, for the given linear multivariable plant

$$\begin{aligned} \dot{x}(t) &= Ax(t) + \sum_{j=1}^{m} B_{*j} u_j(t) \\ y_i(t) &= C_{i*} x(t), \qquad i = 1,\ldots,m \end{aligned} \tag{6.12}$$

where C_{i*} is the ith row of C, B_{*j} is the jth column of B, the relative order is the smallest integer that satisfies the following equation

$$C_{i*}A^{r_{ij}-1}B_{*j} \neq 0 \tag{6.13}$$

Also, the relative order for each element of the transfer function matrix is equal to its relative degree.

6.3 Control Configuration Selection for the Affine Nonlinear Multivariable Plants

In this section, the control configuration analysis for the affine nonlinear multivariable plants based on the plant relative order (degree) is represented. In this approach, a direct graph (*digraph*) using the relative orders of the plant is presented and it will be used as a *structural interaction measure*. In addition, using these relative orders a matrix is provided which will allow us to quantify the structural interaction among input and output variables. This section follows the derivations in (Daoutidis and Kravaris 1992).

6.3.1 Relative Order as an Overall Measure of Control Configuration

Consider an affine nonlinear plant described by equations (6.3). Tools from graph theory are used to represent the nonlinear plant. These tools facilitate the *overall* control configuration analysis of nonlinear plants. Specifically, a digraph representation of the plant using a set of nodes (vertex) and a set of edges is derived. Initially, the basic definitions are given. The vertex and edges in a digraph are defined as follows:

- The *node* or *vertex set* consists of the set of inputs $u_j,\ \ j = 1,\ldots,m$, the set of states $x_k,\ \ k = 1,\ldots,n$ and the set of outputs $y_i,\ \ i = 1,\ldots,m$.
- The *set of edges* contains the direct lines between the nodes according to the following rules:
  - If $\dfrac{\partial f_l(x)}{\partial x_k} \neq 0,\ \ k,l = 1,2,\ldots,n$, then there is an edge from x_k to x_l.
  - If $\dfrac{\partial h_i(x)}{\partial x_k} \neq 0,\ \ k = 1,2,\ldots,n$, then there is an edge from x_k to y_i.
  - If $g_{lj}(x) \neq 0,\ \ l = 1,2,\ldots,n$, then there is an edge from u_j to x_l.

  where $f_l(x)$ and $g_{lj}(x)$ are the lth element of $f(x)$ and $g_j(x)$, respectively.
- Also, a *path* is defined as a directed sequence of several edges, such that the initial vertex of the succeeding edge is the final vertex of the preceding edge. The *length* of the path is defined as the number of edges contained in the path.

Example 6.3.1

Consider the following nonlinear multivariable plant

$$\begin{aligned}\dot{x}_1 &= f_1(x_2, x_3) + g_1(x)u \\ \dot{x}_2 &= f_2(x_1, x_2, x_3) \\ \dot{x}_3 &= f_3(x_1, x_2) \\ y &= h(x_2)\end{aligned}$$

Using the above rules, the corresponding digraph is shown in figure 6.1.

■

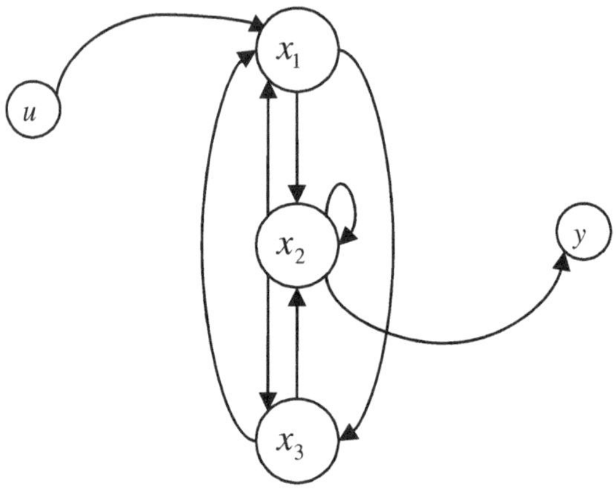

Fig. 6.1 The digraph of example 6.3.1.

Note that the plant digraph shows the pattern of interdependence among the plant variables, and it presents an overall view of the variables relations and interactions. It contains only the general information of the plant dynamics and does not give any detailed information, as provided by the state space model of (6.3). Information such as the exact functional dependence of the vector field $f(x)$ on the state vector x or the numerical values of the plant parameters are not given in the digraph.

There is a motivating and constructive relation between the relative order and the length of paths in the digraph of a nonlinear plant. This is summarized in the following theorem. In the subsequent arguments the term *generic* is used to indicate that the result holds for *nearly all* vector and scalar fields. The interested reader should refer to (Daoutidis and Kravaris 1992) for a detailed proof.

Theorem 6.1

Consider the nonlinear plant given by equations (6.3) and its corresponding digraph. Let l_{ij} denote the length of the shortest path connecting u_j and y_i. Also, let r_{ij} be the relative order between u_j and y_i. Then, $r_{ij} = l_{ij} - 1$ generically.

■

Example 6.3.2

The relative order of the nonlinear plant in example 6.3.1 is $r = 2$ and the length of the shortest path between input and output variables is 3.

■

Note 1. Using theorem 6.1, generic calculation of the relative orders of a nonlinear plant only requires the corresponding digraph of the plant, which is the minimum needed information about the plant. That is, detailed state space information is not necessary for structural control configuration analysis of complex nonlinear multivariable plants.

Note 2. Relative orders are a meaningful measure of the direct influence of an input variable on an output variable.

Note 3. In linear systems, the existence of relative order r_{ij} corresponds to the property of controllability of an output node y_i form the input node u_j.

Relative order is used to measure the *sluggishness* of the plant response. Following theorem stated without proof, shows the relation between largeness of the relative order and the sluggishness of the plant time response (Daoutidis and Kravaris 1992).

Theorem 6.2

Consider the nonlinear plant described by the equations (6.3) at an initial condition $x(0) = x_0$, where x_0 is the nominal steady state. Also, let r_{ij} be the relative order of the output y_i with respect to the manipulated input u_j. Then, the initial response of the output y_i under a unit step change at the input u_j can be approximated, for small times t by

$$y_i(t) \cong L_{g_j} L_f^{r_{ij}-1} h_i(x_0) \frac{t^{r_{ij}}}{r_{ij}!} \tag{6.14}$$

■

The result of the above theorem can be applied to linear time invariant plants given by equations (6.12). With a relative order equal to r_{ij}, the initial response can be approximated as

$$y_i(t) \cong (C_{i*}A^{r_{ij}-1}B_{*j})\frac{t^{r_{ij}}}{r_{ij}!} \tag{6.15}$$

where $y_i(t)$ shows the initial time response of the plant corresponding to a unit step in input $u_j(t)$.

Now, we can analyze the sluggishness of the plant response using the result of theorem 6.2. Using equation (6.14), the initial slope of the plant response is characterized in the three following cases and is shown in figure 6.2.

- $r_{ij} = 1$, in this case the initial slope of the response is not zero.
- $r_{ij} = 2$, in this case the initial slope of the response is zero but its rate of change is not zero.
- $r_{ij} > 2$, in this case the initial slope of the response and its rate of change is zero.

Example 6.3.3

Consider the cascade of two continuous stirred tank reactors (CSTR) as shown in figure 6.3 with the following nonlinear state space equations. Where c_p is the heat capacity, E is the activation energy, F is the volumetric flow rate, Q_1, Q_2 are the heat inputs to tanks 1 and 2, T_1, T_2 are the temperatures in tanks 1 and 2, T_0 is the inlet temperature, V is the volume, $-\Delta H$ is the heat of reaction,

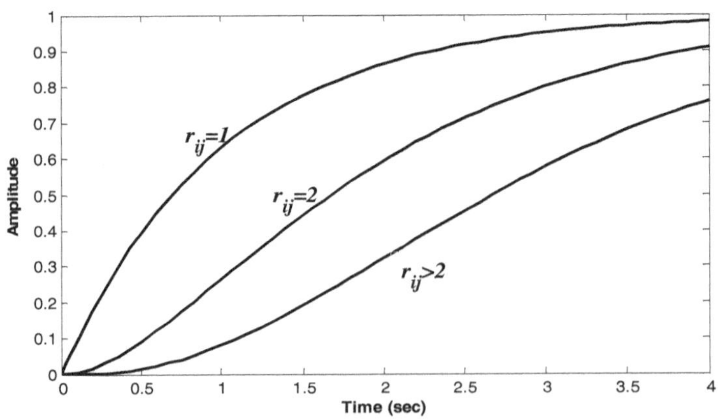

Fig. 6.2 Relative order and sluggishness of the corresponding plant time responses.

c_{A1}, c_{A2} are the molar concentration in tanks 1 and 2, c_{A0} is the inlet molar concentration, k_0 is the Arrhenius frequency factor, ρ is the density, $y = c_{A2}$ is the

$$\frac{dc_{A1}}{dt} = \tfrac{F}{V}(c_{A0} - c_{A1}) - k_0 e^{-E/RT_1} c_{A1}^2$$
$$\frac{dc_{A2}}{dt} = \tfrac{F}{V}(c_{A1} - c_{A2}) - k_0 e^{-E/RT_2} c_{A2}^2$$
$$\frac{dT_1}{dt} = \tfrac{F}{V}(T_0 - T_1) + \tfrac{-\Delta H}{\rho c_p} k_0 e^{-E/RT_1} c_{A1}^2 + \tfrac{1}{V\rho c_p} Q_1$$
$$\frac{dT_2}{dt} = \tfrac{F}{V}(T_1 - T_2) + \tfrac{-\Delta H}{\rho c_p} k_0 e^{-E/RT_2} c_{A2}^2 + \tfrac{1}{V\rho c_p} Q_2$$

process output, and $u_1 = Q_1$, $u_2 = Q_2$ are the process inputs. The corresponding digraph is as shown in figure 6.4.

Figure 6.4 shows that the relative orders corresponding to $(Q_1 - y)$ and $(Q_2 - y)$ are equal to 3, $r_{11} = 3$, and 2, $r_{12} = 2$, respectively. Therefore, it is obvious that for the above multi input single output plant, Q_2 as one of the manipulated variables has a more direct effect or physically closeness to the main output y. So, Q_2 will be the manipulated input to control the output and to access the good control quality characteristics.

■

According to the above theorems and examples, relative order can be used as a measure to evaluate the structural interaction or the dependence among the plant variables. To formulate the above analysis and to study the structural interaction of the multivariable plants, the following definition is used.

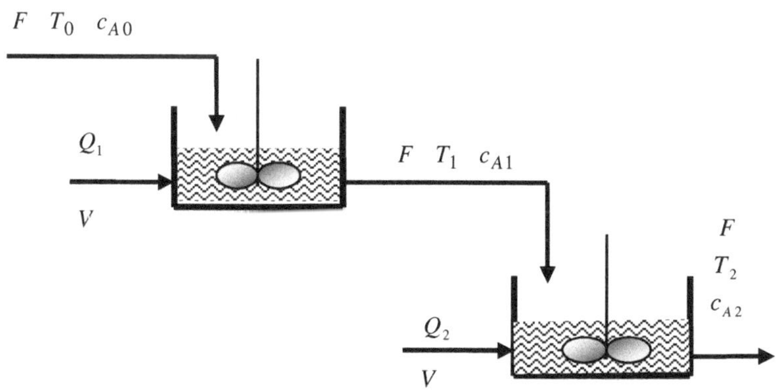

Fig. 6.3 A cascade of two continuous stirred tank reactors (CSTR).

Definition 6.4. Consider the nonlinear multivariable plant described by equations (6.3). The *relative-order matrix* of the plant is defined as

$$M = \begin{bmatrix} r_{11} & \cdots & r_{1m} \\ \vdots & \ddots & \vdots \\ r_{m1} & \cdots & r_{mm} \end{bmatrix} \tag{6.16}$$

where r_{ij} is the relative order between the manipulated variable u_j, and the measured variable y_i. It is obvious that the relative order matrix is a numerical representation of the corresponding digraph of the nonlinear multivariable plant.

6.3.2 Evaluation of the Control Configuration

In section 6.3.1, the digraph and relative order matrix are introduced as an overall analysis of control configurations. In this section, a key theorem is presented that is used for control configuration and appropriate input-output pair selection of nonlinear multivariable plants.

Definition 6.5. A *structural matrix* is a matrix which only keeps the zero and non-zero information of a matrix. In an equivalent structural matrix, fixed zeros of the original matrix is kept and nonzero elements are replaced with arbitrary nonzero parameters.

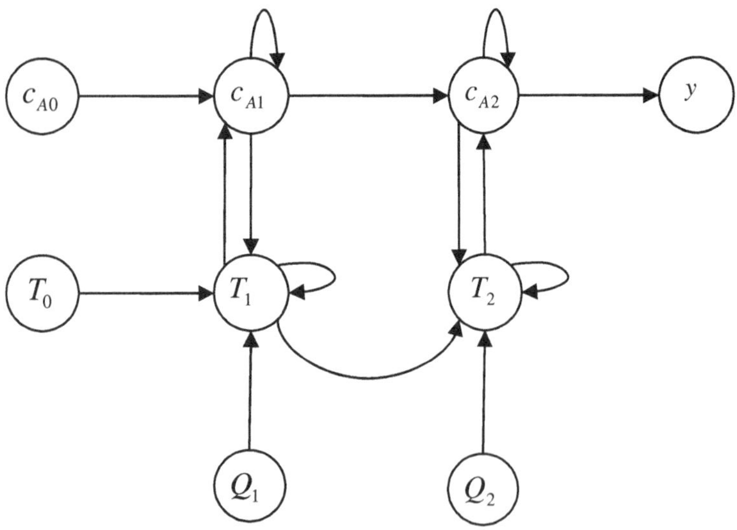

Fig. 6.4 The digraph of CSTR.

Definition 6.6. The *generic rank* of a structural matrix is its maximal possible rank when its nonzero elements get numerical values.

Theorem 6.3

Consider the nonlinear plant described by equations (6.3) and define its characteristic matrix $C(x)$ as

$$C(x)=\begin{bmatrix} L_{g_1}L_f^{r_1-1}h_1(x) & \cdots & L_{g_m}L_f^{r_1-1}h_1(x) \\ \vdots & & \vdots \\ L_{g_1}L_f^{r_m-1}h_m(x) & \cdots & L_{g_m}L_f^{r_m-1}h_m(x) \end{bmatrix} \tag{6.17}$$

where $r_i=\min\{r_{ij}\}$, $j=1,2,\ldots,m$. Then, the generic rank of the structural matrix equivalent to $C(x)$ will be m, if and only if the outputs can be rearranged so that the minimum relative order in each row of the relative-order matrix appears in the major diagonal position, i.e. this new matrix denoted by M_r takes the form

$$M_r=\begin{bmatrix} r_1 & r_{12} & & r_{1m} \\ r_{21} & r_2 & & r_{2m} \\ \vdots & & \ddots & \vdots \\ r_{m1} & r_{m2} & & r_m \end{bmatrix} \tag{6.18}$$

Proof. The theorem is proved by contradiction using the derivation in (Daoutidis and Kravaris 1992). The *only if part*: Suppose that the structural matrix equivalent to $C(x)$ has generic rank equal to m and the output rearrangement is not possible. This implies that there is at least one input u_{j^*} for which one of the following two is true:

1. There is no output y_i with the minimum relative order at the j^{*th} column of the relative order matrix M_r, i.e. there is no output y_i such that $r_i=r_{ij^*}$.
2. There are two or more outputs, e.g. y_{i1} and y_{i2}, with minimum relative order appear at the j^{*th} column of the relative-order matrix M_r and nowhere else, i.e. $r_{i1}=r_{i1j^*}$, $r_{i2}=r_{i2j^*}$ and $r_{i1j}>r_{i1j^*}$, $r_{i2j}>r_{i2j^*}$ for $j\neq j^*$.

In the first case, we would have

$$L_{g_{j^*}}L_f^{r_i-1}h_i(x_0)=0$$

for every i, and therefore the j^{*th} column of the characteristic matrix (and its structural equivalent) would be zero. In the second case, we would have

$$L_{g_{j^*}} L_f^{r_{i1}-1} h_{i1}(x_0) \neq 0,\ \ L_{g_{j^*}} L_f^{r_{i2}-1} h_{i2}(x_0) \neq 0$$

and

$$L_{g_j} L_f^{r_{i1j}-1} h_{i1}(x_0) = 0,\ \ L_{g_j} L_f^{r_{i2j}-1} h_{i2}(x_0) = 0$$

for every $j \neq j^*$. But then, corresponding to the outputs y_{i1} and y_{i2}, rows of the characteristic matrix would have only one nonzero element, at the same position (the j^{*th}). In both cases, a rank deficiency would result, contrary to our assumption. Therefore, by contradiction, the suggested output rearrangement is always possible.

The *if part*: Suppose that the suggested output rearrangement is possible, but the structural matrix equivalent to $C(x)$ has rank deficiency. This implies either of the following for this matrix:

1. At least one row or column has zeros in all positions.
2. There are $k(k > 2)$ columns or rows that cause the rank deficiency in a non-trivial way.

In the first case, all the relative orders in a row or column are equal to infinity. In the second case, in order that the rank deficiency may exist, we must have at least $m-(k-1)$ zeros at the same positions in all k columns or rows. This leaves at most $(k-1)$ nonzero elements at the same $(k-1)$ positions of all k rows or columns. However, because of the rearrangement, there should be k nonzero elements in the diagonal positions of these k rows or columns, i.e. in k distinct positions. In both cases, the contradiction is clear, and the theorem is proved.

■

Note. The output rearrangement indicated in the above theorem has similar implications as the permutation of rows in the RGA analysis of transfer function matrices for linear multivariable plants.

According to the above theorem, the following points are valid for a nonlinear multivariable plant with nonsingular characteristic matrix $C(x)$, where the equivalent structural matrix has full generic rank:

- The suggested output rearrangement indicates the input-output pairing $(u_i - y_i)$ corresponding to on-diagonal elements and diagonal dominant structural interaction.

- After the output rearrangement, the off-diagonal relative orders allow the evaluation of structural interaction between a specific input-output pair and the remaining input and output variables.
- The difference between off-diagonal and diagonal elements in rows $\sum_{j \neq i} r_{ij} - r_i$, or columns $\sum_{j \neq i} r_{ji} - r_i$, of the relative order matrix can be used as an overall measure of structural interaction between system inputs-outputs. Larger differences indicate the weaker structural interaction among the manipulated and measured variables. Also, this property can be used as a tool to choose the groups of inputs and outputs.
- In the extreme case of a decoupled nonlinear multivariable plant, the decoupled matrix M_r will be

$$M_r = \begin{bmatrix} r_1 & \infty & & \infty \\ \infty & r_2 & & \infty \\ \vdots & & \ddots & \vdots \\ \infty & \infty & & r_m \end{bmatrix} \tag{6.19}$$

6.3.3 *Input-Output Pairing Rule using the Relative-Order Matrix*

The following rules are proposed to select the appropriate control configuration or input-output pair for the affine nonlinear multivariable plants:

- Compute the characteristic matrix corresponding to the nonlinear multivariable plant. If this matrix is full rank, decentralized control configuration is feasible.
- Compute the relative order-matrix equivalent to the characteristic matrix and using the outputs rearrangement (or rows permutation) assign the minimum relative orders of its rows on the major diagonal. These diagonal elements characterize the appropriate input-output pair.

Example 6.3.4

Consider a single-effect evaporator as shown in figure 6.5. A solution stream at solute molar concentration x_F enters the evaporator at a molar flow rate F. Heat provided by steam Q is used to vaporize the water, producing a vapor stream D and a liquid effluent B at a solute concentration x_B. Details of the model and assumptions can be found in (Daoutidis and Kravaris 1992). Manipulating the equations governing the process, the following state space equations of the plant are derived

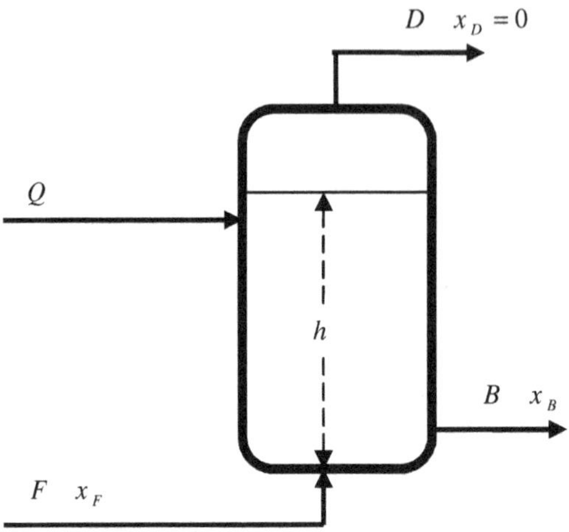

Fig. 6.5 A single-effect evaporator.

$$\dot{x} = \begin{bmatrix} 0 \\ -\dfrac{B_s x_2}{Ac(x_1 + h_s)} \end{bmatrix} + \begin{bmatrix} -\dfrac{1}{Ac} & -\dfrac{1}{\Delta H_v Ac} \\ 0 & \dfrac{x_2 + x_{Bs}}{\Delta H_v Ac(x_1 + h_s)} \end{bmatrix} \begin{bmatrix} u_1 \\ u_2 \end{bmatrix} + \begin{bmatrix} 0 \\ \dfrac{F}{Ac(x_1 + h_s)} \end{bmatrix} d$$

$$y = \begin{bmatrix} y_1 \\ y_2 \end{bmatrix} = \begin{bmatrix} x_1 \\ x_2 \end{bmatrix}$$

where $x_1 = h - h_s$, $x_2 = x_B - x_{Bs}$, $u_1 = B - B_s$, $u_2 = Q - Q_s$, $d = x_F - x_{Fs}$ the subscript s denotes the nominal steady-state value and d is the disturbance input. Also, A denotes the cross-section area, F, B and D indicate the molar flow rates and in the steady-state $F_s = B_s + D_s$, c denotes the molar density of the feed and bottom streams, h is the liquid level in the evaporator and x_F, x_B indicate the solute concentration at the feed and bottom streams, respectively (in mole fractions). Also, ΔH_v is the latent heat of the vaporization and Q is the heat input to the evaporator.

According to the above nonlinear state space model, the corresponding characteristic matrix, $C(x)$ is

$$C(x) = \begin{bmatrix} L_{g_1} h_1(x) & L_{g_2} h_1(x) \\ L_{g_1} h_2(x) & L_{g_2} h_2(x) \end{bmatrix} = \begin{bmatrix} -\dfrac{1}{Ac} & -\dfrac{1}{\Delta H_v Ac} \\ 0 & \dfrac{x_2 + x_{Bs}}{\Delta H_v Ac(x_1 + h_s)} \end{bmatrix}$$

$C(x)$ is nonsingular and according to theorem 6.3, the relative-order matrix is

$$M_r = \begin{bmatrix} 1 & 1 \\ 2 & 1 \end{bmatrix}$$

and it proposes the $(u_1 - y_1, u_2 - y_2)$ as the appropriate input-output pair.

■

Example 6.3.5

Consider the CSTR (continuous stirred tank reactor) shown in figure 6.6. In this process, two solution streams consisting of species *A* and *B*, at volumetric flow rates F_A and F_B, temperatures T_A and T_B, and concentrations c_{A0} and c_{B0}, enter the reactor where the elementary reaction $A + B \rightarrow C + D$ takes place. The effluent stream leaves the reactor at a flow rate F, concentrations c_A, c_B, c_C, c_D and temperature T. Heat may be added or removed from the system at a rate *Q*, using an appropriate heating/cooling system. Also, assuming constant density ρ and constant heat capacity c_p for the liquid streams and neglecting the heat of solution effects, the material and energy balances equations give a nonlinear state space model of the process (Daoutidis and Kravaris 1992). Let

$$x_1 = V - V_s,\ x_2 = c_A - c_{As},\ x_3 = c_B - c_{Bs},\ x_4 = c_C - c_{Cs},\ x_5 = T - T_s$$

and

$$u_1 = F_A - F_{As},\ u_2 = F_B - F_{Bs},\ u_3 = F - F_s,\ u_4 = Q - Q_s$$

also

$$y_1 = x_1,\ y_2 = x_2,\ y_3 = x_4,\ y_4 = x_5$$

where the subscript " s " denotes a nominal steady state value. Then, the nonlinear state space model is as follows

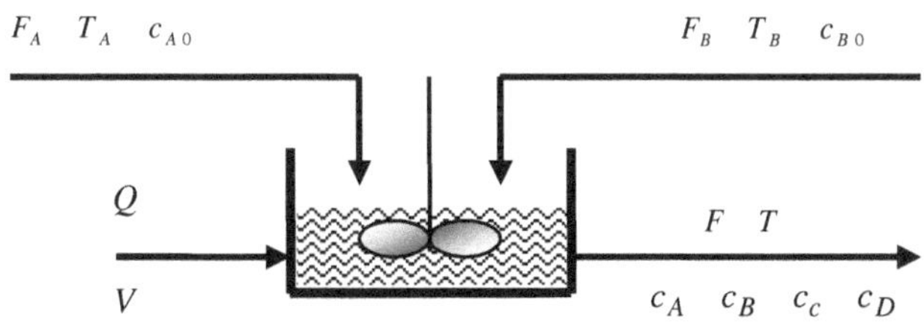

Fig. 6.6 A Continuous Stirred Tank Reactor (CSTR).

$$\dot{x}_1 = u_1 + u_2 - u_3$$

$$\dot{x}_2 = \frac{F_{As}(c_{A0} - x_2 - c_{As}) - F_{Bs}(x_2 + c_{As})}{x_1 + V_s} - k(x_2 + c_{As})(x_3 + c_{Bs})e^{-E/R(x_5+T_s)}$$
$$+ \frac{u_1(c_{A0} - x_2 - c_{As}) - u_2(x_2 + c_{As})}{x_1 + V_s}$$

$$\dot{x}_3 = \frac{F_{Bs}(c_{B0} - x_3 - c_{Bs}) - F_{As}(x_3 + c_{Bs})}{x_1 + V_s} - k(x_2 + c_{As})(x_3 + c_{Bs})e^{-E/R(x_5+T_s)}$$
$$+ \frac{u_2(c_{B0} - x_3 - c_{Bs}) - u_1(x_3 + c_{Bs})}{x_1 + V_s}$$

$$\dot{x}_4 = -\frac{(F_{As} + F_{Bs})(x_4 + c_{cs})}{x_1 + V_s} + k(x_2 + c_{As})(x_3 + c_{Bs})e^{-E/R(x_5+T_s)} - \frac{(x_4 + c_{cs})(u_1 + u_2)}{x_1 + V_s}$$

$$\dot{x}_5 = \frac{(T_A - x_5 - T_s)F_{As} + (T_B - x_5 - T_s)F_{Bs}}{x_1 + V_s} + \frac{Q_s}{\rho c_p(x_1 + V_s)}$$
$$- \frac{k\,\Delta H}{\rho c_p}(x_2 + c_{As})(x_3 + c_{Bs})e^{-E/R(x_5+T_s)}$$
$$+ \frac{(T_A - x_5 - T_s)u_1 + (T_B - x_5 - T_s)u_2}{x_1 + V_s} + \frac{u_4}{\rho c_p(x_1 + V_s)}$$

where V and $(-\Delta H)$ denote the volume and the heat of reaction, respectively. So, using equations (6.11) and (6.16), the corresponding relative-order matrix to evaluate the structural interaction is

$$M = \begin{bmatrix} r_{11} & r_{12} & r_{13} & r_{14} \\ r_{21} & r_{22} & r_{23} & r_{24} \\ r_{31} & r_{32} & r_{33} & r_{34} \\ r_{41} & r_{42} & r_{43} & r_{44} \end{bmatrix} = \begin{bmatrix} 1 & 1 & 1 & \infty \\ 1 & 1 & 2 & 2 \\ 1 & 1 & 2 & 2 \\ 1 & 1 & 2 & 1 \end{bmatrix}$$

also the characteristic matrix is

$$C(x) = \begin{bmatrix} L_{g_1}h_1(x) & L_{g_2}h_1(x) & L_{g_3}h_1(x) & 0 \\ L_{g_1}h_2(x) & L_{g_2}h_2(x) & 0 & 0 \\ L_{g_1}h_3(x) & L_{g_2}h_3(x) & 0 & 0 \\ L_{g_1}h_4(x) & L_{g_2}h_4(x) & 0 & L_{g_4}h_4(x) \end{bmatrix}$$

where the characteristic matrix and its equivalent structural matrix have full generic rank. According to the proposed pairing rule, using the output rearrangement, by permuting the 1st and 3rd rows of the relative-order matrix, we obtain the following matrix which satisfies the pairing conditions

$$M_r = \begin{bmatrix} 1 & 1 & 2 & 2 \\ 1 & 1 & 2 & 2 \\ 1 & 1 & 1 & \infty \\ 1 & 1 & 2 & 1 \end{bmatrix}$$

and it indicates the following appropriate input-output pair for decentralized control of the process $(u_1 - y_3, u_2 - y_2, u_3 - y_1, u_4 - y_4)$.

■

Example 6.3.6

Consider the Quadruple-tank introduced in example 2.3.1. Using the state variables $x_1 = h_1$, $x_2 = h_2$, $x_3 = h_3$, $x_4 = h_4$ and the manipulated variables $u_1 = v_1$, $u_2 = v_2$ and the measured variables $y_1 = x_1$, $y_2 = x_2$, the nonlinear state space model of the Quadruple-tank is

$$\dot{x}_1 = -\frac{a_1}{A_1}\sqrt{2gx_1} + \frac{a_3}{A_1}\sqrt{2gx_3} + \frac{\gamma_1 k_1}{A_1}u_1$$

$$\dot{x}_2 = -\frac{a_2}{A_2}\sqrt{2gx_2} + \frac{a_4}{A_2}\sqrt{2gx_4} + \frac{\gamma_2 k_2}{A_2}u_2$$

$$\dot{x}_3 = -\frac{a_3}{A_3}\sqrt{2gx_3} + \frac{(1-\gamma_2)k_2}{A_3}u_2$$

$$\dot{x}_4 = -\frac{a_4}{A_4}\sqrt{2gx_4} + \frac{(1-\gamma_1)k_1}{A_4}u_1$$

The digraph of the plant is shown in figure 6.7 and the relative-order matrix is given below

$$M = \begin{bmatrix} r_{11} & r_{12} \\ r_{21} & r_{22} \end{bmatrix} = \begin{bmatrix} 1 & 2 \\ 2 & 1 \end{bmatrix} = M_r$$

also the characteristic matrix of the Quadruple-tank is

$$C(x) - \begin{bmatrix} L_{g_1}h_1(x) & L_{g_2}h_1(x) \\ L_{g_1}h_2(x) & L_{g_2}h_2(x) \end{bmatrix} = \begin{bmatrix} \frac{\gamma_1 k_1}{A_1} & 0 \\ 0 & \frac{\gamma_2 k_2}{A_2} \end{bmatrix} \tag{6.20}$$

It is obvious that the characteristic matrix is generically full rank. Hence, using the result of theorem 6.3, the appropriate input-output pair is $(u_1 - y_1, u_2 - y_2)$.

■

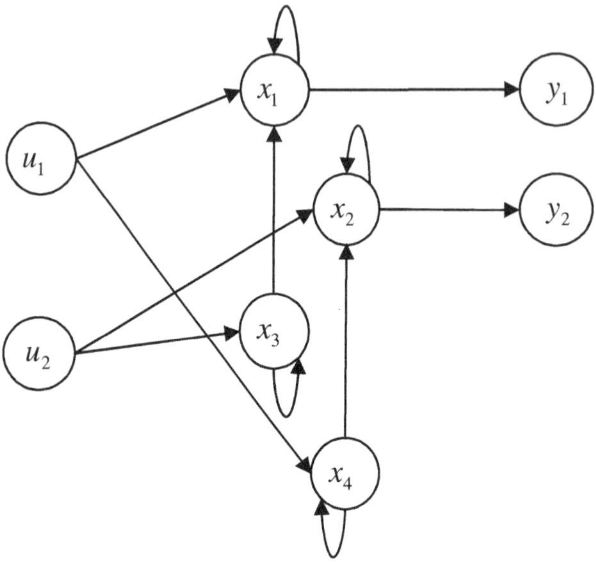

Fig. 6.7 The digraph of Quadruple-tank.

6.4 The Nonlinear RGA

In section 6.3, we presented the control configuration selection for nonlinear multivariable plants based on the relative-order matrix. This structural interaction measure provides an overview of the plant variables coupling but is unable to show the effects of plant characteristics on its interaction.

The RGA introduced in chapter 2 is the most widely used method to input-output pairing for linear multivariable plants. In this section, we introduce an extension of the RGA for nonlinear plants based on (Glad 1999) in section 6.4.1 and the nonlinear RGA (NRGA) applicable to nonlinear plants based on (Moaveni and Khaki-Sedigh 2007) in section 6.4.2. Both of these ideas are in a way extensions of the classical RGA to generate pairing methods for nonlinear multivariable plants.

6.4.1 An Extension of the RGA for Nonlinear Multivariable Plants

Consider a nonlinear multivariable plant with its input-output model as

$$y = H(u) \tag{6.21}$$

where u and y are the input and output m-vectors respectively and H is a vector field from $\Re^m$ to $\Re^m$. Also, we assume that for each constant input there exists an equilibrium point and a constant output. Hence, for a constant reference input u^0 the output is $y^0 = H(u^0)$.

Similar to the relative gain derivation in chapter 2, introduce the following nonlinear static relationship

$$y_i = \phi_{ij}(u_j) = H_i(u_1^0,\dots,u_{j-1}^0,u_j,u_{j+1}^0,\dots,u_m^0) \tag{6.22}$$

where H_i denotes the ith component of H and ϕ_{ij} shows the y_i dependence on u_j. Also, if H is invertible we have

$$u_j = \psi_{ji}(y_i) = (H)^{-1}{}_i(y_1^0,\dots,y_{i-1}^0,y_i,y_{j+1}^0,\dots,y_m^0) \tag{6.23}$$

where ψ_{ji} indicates the u_j dependence on y_i, when all other outputs are kept at the nominal values y_k^0, $k \neq i$. So, the *nonlinear static RGA* can be defined as follows (Glad 1999).

Definition 6.7 For a steady-state input-output model as (6.21), where H is invertible, the nonlinear static Relative Gain is defined as

$$\lambda_{nl-ij} = [\Lambda_{nl}]_{ij} = \phi_{ij}\left(\psi_{ji}(y_i)\right) \tag{6.24}$$

Note that in the above definition, we need the input-output model of the nonlinear plant and the vector function H should be invertible. Consider the nonlinear multivariable plant described by the affine state space model

$$\begin{aligned} \dot{x}(t) &= f\left(x(t)\right) + g\left(x(t)\right)u(t) \\ y(t) &= h\left(x(t)\right) \end{aligned} \tag{6.25}$$

where $x \in \Re^n$, $u \in \Re^m$ and $y \in \Re^m$ denote the state, input and output vectors, respectively. It is desired to extend the RGA definition to the class of affine nonlinear plants. The characteristic matrix with vector relative order $\{r_1,\dots,r_m\}$ given by equation (6.17) and the following vectors are assumed

$$a(x) = \begin{bmatrix} L_f^{r_1}h_1 \\ \vdots \\ L_f^{r_m}h_m \end{bmatrix}, \quad Y = \begin{bmatrix} y_1^{(r_1)} \\ \vdots \\ y_m^{(r_m)} \end{bmatrix} \tag{6.26}$$

Define the following new coordinates to obtain the *normal form* of nonlinear plant

$$\xi^i = \begin{bmatrix} \xi_1^i \\ \xi_2^i \\ \vdots \\ \xi_{r_i}^i \end{bmatrix} = \begin{bmatrix} y_i \\ \dot{y}_i \\ \vdots \\ y_i^{(r_i-1)} \end{bmatrix}, \quad i = 1,\ldots,m$$
$$\eta = \begin{bmatrix} \eta_1 \\ \vdots \\ \eta_{n-r} \end{bmatrix} = \begin{bmatrix} \varphi_{r+1}(x) \\ \vdots \\ \varphi_n(x) \end{bmatrix} \tag{6.27}$$

where $r = r_1 + r_2 + \cdots + r_m \le n$ and functions $\varphi_k(x)$, $k = r+1,\ldots,n$ can be computed by solving the equations $L_{g_j}\varphi_k(x) = 0$ for all $j = 1,\ldots,m$ (Isidori 1995). Then

$$\begin{cases} \dot{\xi}_1^i = \xi_2^i \\ \quad\vdots \\ \dot{\xi}_{r_i}^i = a_i(\xi,\eta) + \sum_{j=1}^{m} [C(\xi,\eta)]_{ij}\, u_j \\ \dot{\eta} = b(\xi,\eta) + d(\xi,\eta)u \end{cases}, \quad i = 1,\ldots,m \tag{6.28}$$
$$y_i = \xi_1^i, \quad i = 1,\ldots,m$$

Hence, using equations (6.26) and (6.28), the plant input-output relations can be written as

$$Y = a(\xi,\eta) + C(\xi,\eta)u \tag{6.29}$$

The following theorem is now presented to evaluate the interaction in a nonlinear multivariable plant.

Theorem 6.4

For a nonlinear multivariable plant given by equation (6.25), the operator $\tilde{y} = \Lambda_{nl-ij}(y)$ given by equation (6.24) can be represented by equations (6.28), where u_j can be calculated as

$$\begin{aligned} u_j &= [C^{-1}(\xi_{Yi},\eta)\left(Y_i - a(\xi_{Yi},\eta)\right)]_j \\ \dot{\eta} &= b(\xi_{Yi},\eta) + d(\xi_{Yi},\eta)C^{-1}(\xi_{Yi},\eta)\left(Y_i - a(\xi_{Yi},\eta)\right) \end{aligned} \tag{6.30}$$

where Y_i denotes a vector of zeros with $y_i^{(r_i)}$ as its ith element and ξ_{Yi} is the vector obtained when ξ_k^i, $k = 1,\ldots,r_i$ is replaced by $y_i^{(k-1)}$, and all the other components are zero.

Proof. It is obvious that equations $\xi_{r_i}^{i} = a_i(\xi,\eta) + \sum_{j=1}^{m} [C(\xi,\eta)]_{ij} u_j$, $i = 1,\ldots,m$ in equation (6.28) can be represented in the vector form as

$$\underbrace{\begin{bmatrix} \xi_{r_1}^{1} \\ \xi_{r_2}^{2} \\ \vdots \\ \xi_{r_m}^{m} \end{bmatrix}}_{\xi_{r_i}, i=1,\ldots,m} = \underbrace{\begin{bmatrix} a_1(\xi,\eta) \\ a_2(\xi,\eta) \\ \vdots \\ a_m(\xi,\eta) \end{bmatrix}}_{a(\xi,\eta)} + [C(\xi,\eta)] \underbrace{\begin{bmatrix} u_1 \\ u_2 \\ \vdots \\ u_m \end{bmatrix}}_{u} = Y \tag{6.31}$$

using equation (6.29), the control law is

$$u = C^{-1}(\xi,\eta)(Y - a(\xi,\eta)) \tag{6.32}$$

Hence, to analyze the effect of the inputs on y_i when all other outputs are under tight control equation (6.32) can be rewritten as

$$u = C^{-1}(\xi_{Yi},\eta)(Y_i - a(\xi_{Yi},\eta)) \tag{6.33}$$

Substituting equation (6.33) in equation (6.28) will keep $\xi_k^i = 0\ k \neq i$, provided that their initial conditions are zero. This consequently keeps $y_k = 0\ k \neq i$, while we get $\xi_j^i = y_i^{(j)}$. Substituting the jth component of the input vector

$$u_j = [C^{-1}(\xi_{Yi},\eta)(Y_i - a(\xi_{Yi},\eta))]_j \tag{6.34}$$

into equation (6.28) and zeroing the other components of u gives the desired result.

■

The above theorem introduces a relationship between the $u_j - y_i$ loop and the other control loops. Substituting u_j from equation (6.34) into equation (6.28), gives

$$\tilde{y}_i^{(r_i)} = [C(\xi,\eta)]_{ij} [C(\xi_{Yi},\varsigma)]_{ji}^{-1} y_i^{(r_i)} + v \tag{6.35}$$

This is a direct link between $\tilde{y}$ and y, where v stands for all the remaining parts of the equation. Equation (6.35) introduces a matrix to evaluate the control configuration in the nonlinear plant. It is an extension of the RGA for nonlinear multivariable plants and can be rewritten as

$$\Lambda_{nl} = C(x).*C(x)^{-T} \tag{6.36}$$

where ".*" denotes the element-by-element product.

Example 6.4.1

Consider the Quadruple-tank introduced in example 6.3.6 and the corresponding characteristic matrix given by equation (6.20). Using equation (6.36), the nonlinear extension of RGA for the Quadruple-tank is

$$\Lambda_{nl} = C(x).*C(x)^{-T} = \begin{bmatrix} \frac{\gamma_1 k_1}{A_1} & 0 \\ 0 & \frac{\gamma_2 k_2}{A_2} \end{bmatrix} .* \begin{bmatrix} \frac{\gamma_1 k_1}{A_1} & 0 \\ 0 & \frac{\gamma_2 k_2}{A_2} \end{bmatrix}^{-T} = \begin{bmatrix} 1 & 0 \\ 0 & 1 \end{bmatrix}$$

Thus, the proposed appropriate input-output pair is $(u_1 - y_1, u_2 - y_2)$. This coincides with the previous result obtained by the relative-order matrix in example 6.3.6. However, in example 2.3.1 the RGA analysis showed the input-output pairing depends on the parameters γ_1 and γ_2. This result is not deducible from the above methodology. Hence, to further analyze the results we study the closed loop step response performances under decentralized control. Consider the following physical constants

$$\begin{aligned}
&A_1 = A_3 = 28 \quad (cm^2) \\
&A_2 = A_4 = 32 \quad (cm^2) \\
&a_1 = a_3 = 0.071 \quad (cm^2) \\
&a_2 = a_4 = 0.057 \quad (cm^2) \\
&g = 981 \quad (cm/s^2) \\
&k_i = 2.9
\end{aligned}$$

and set the following conditions for the Quadruple-tank process

$$\begin{aligned}
&h_1 = 12.7 \quad (cm), \quad h_2 = 12.4 \quad (cm) \\
&h_3 = 7.4 \quad (cm), \quad h_4 = 8.7 \quad (cm) \\
&\gamma_1 = 0.3, \quad \gamma_2 = 0.4
\end{aligned}$$

For these conditions $\gamma_1 + \gamma_2 < 1$, the RGA proposes $(u_1 - y_2, u_2 - y_1)$ as an appropriate input-output pair, and in the above method $(u_1 - y_1, u_2 - y_2)$ is the proposed pair. Using the decentralized control structure with PI controllers introduced by Johansson (2000) the closed loop step responses of the plant for these two input-output pairing are shown in figure 6.8. It is obvious that the closed loop performance corresponding to the RGA pairing $(u_1 - y_2, u_2 - y_1)$ is superior.

■

This example shows the necessity for serious improvement in the input-output pairing of nonlinear multivariable plant.

6.4.2 The Nonlinear-RGA for Affine Nonlinear Multivariable Plants

In this section, a mathematical relationship between the inputs and outputs of the affine nonlinear multivariable plants is derived. Employing this relationship, a new input-output pairing analysis based on the extended relative gain definition called the nonlinear-RGA (NRGA) will be presented (Moaveni and Khaki-Sedigh 2007).

Consider the following affine nonlinear multivariable plant

$$\begin{aligned} \dot{x}(t) &= f(x(t)) + \sum_{j=1}^{m} g_j(x(t))u_j(t) \\ y_1(t) &= h_1(x(t)) \\ &\vdots \\ y_m(t) &= h_m(x(t)) \end{aligned} \tag{6.37}$$

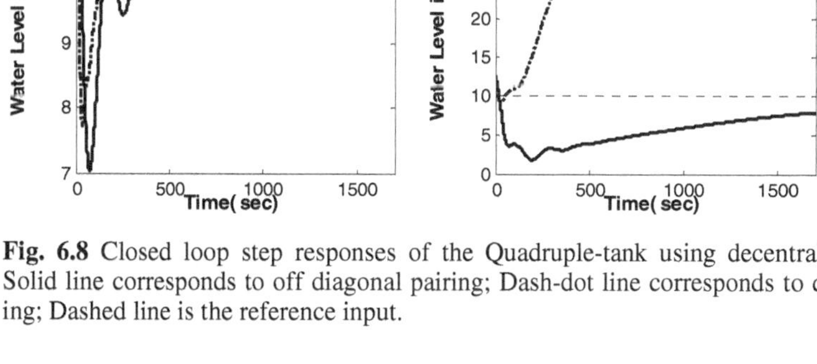

Fig. 6.8 Closed loop step responses of the Quadruple-tank using decentralized control. Solid line corresponds to off diagonal pairing; Dash-dot line corresponds to diagonal pairing; Dashed line is the reference input.

As $h_i(x)$ are assumed smooth functions, the Taylor series expansion of $y_i(t)$ at $t=\tau$ is

$$y_i(t)=\sum_{k=0}^{\infty}\frac{d^k y_i(\tau)}{dt^k}\frac{(t-\tau)^k}{k!} \tag{6.38}$$

where

$$\begin{aligned}\dot{y}_i(t)&=\frac{dy_i}{dt}=\frac{dh_i(x)}{dt}=\frac{dh_i(x)}{dx}\times\frac{dx}{dt} = \frac{dh_i(x)}{dx}\times(f(x)+\sum_{j=1}^{m}g_j(x)u_j)=\\ &=\frac{dh_i(x)}{dx}\times f(x)+\sum_{j=1}^{m}\frac{dh_i(x)}{dx}g_j(x)u_j = L_{f(x)}h_i(x)+\sum_{j=1}^{m}L_{g_j}h_i(x)u_j\end{aligned} \tag{6.39}$$

Hence

$$y_i^{(k)}(t)=L_{f(x)}^k h_i(x)+\sum_{j=1}^{m}L_{g_j(x)}L_{f(x)}^{k-1}h_i(x)u_j \tag{6.40}$$

Therefore, equation (6.38) can be rewritten as

$$y_i(t)=\sum_{k=0}^{\infty}L_{f(x)}^k h_i(x)\frac{(t-\tau)^k}{k!}+\sum_{k=0}^{\infty}\sum_{j=1}^{m}L_{g_j(x)}L_{f(x)}^{k-1}h_i(x)u_j\frac{(t-\tau)^k}{k!} \tag{6.41}$$

Thus for the plant outputs we have

$$\begin{aligned}y_i(t)&=\sum_{k=0}^{\infty}L_{f(x)}^k h_i(x)\frac{(t-\tau)^k}{k!}\\ &+\left[\sum_{k=0}^{\infty}L_{g_1(x)}L_{f(x)}^{k-1}h_i(x)\frac{(t-\tau)^k}{k!}\quad\cdots\quad\sum_{k=0}^{\infty}L_{g_m(x)}L_{f(x)}^{k-1}h_i(x)\frac{(t-\tau)^k}{k!}\right]\begin{bmatrix}u_1(t)\\ \vdots\\ u_m(t)\end{bmatrix}\end{aligned} \tag{6.42}$$

and

$$\begin{aligned}\begin{bmatrix}y_1(t)\\ \vdots\\ y_m(t)\end{bmatrix}&=\sum_{k=0}^{\infty}\begin{bmatrix}L_{f(x)}^k h_1(x)\\ \vdots\\ L_{f(x)}^k h_m(x)\end{bmatrix}\frac{(t-\tau)^k}{k!}\\ &+\underbrace{\sum_{k=0}^{\infty}\frac{(t-\tau)^k}{k!}\begin{bmatrix}L_{g_1(x)}L_{f(x)}^{k-1}h_1(x) & \cdots & L_{g_m(x)}L_{f(x)}^{k-1}h_1(x)\\ \vdots & \ddots & \vdots\\ L_{g_1(x)}L_{f(x)}^{k-1}h_m(x) & \cdots & L_{g_m(x)}L_{f(x)}^{k-1}h_m(x)\end{bmatrix}}_{\Phi}\begin{bmatrix}u_1(t)\\ \vdots\\ u_m(t)\end{bmatrix}\end{aligned} \tag{6.43}$$

Now, to propose an interaction measure or input-output pairing analysis we can use the RG definition of Bristol in chapter 2. The RG is the ratio of the plant gain in an isolated loop to plant gain in the same loop when all other control loops are closed. The plant gain in an isolated loop can be computed as

$$\left.\frac{\Delta y_i}{\Delta u_j}\right|_{\substack{\Delta u_k=0\\ k\neq j}}=\sum_{k=0}^{\infty}\frac{(t-\tau)^k}{k!}L_{g_j(x)}L_{f(x)}^{k-1}h_i(x)=\Phi_{ij} \tag{6.44}$$

Also, the plant gain in that same loop when all other control loops are closed is

$$\left.\frac{\Delta y_i}{\Delta u_j}\right|_{\substack{\Delta y_k=0\\ k\neq i}}=\frac{1}{[\Phi^{-1}]_{ji}} \tag{6.45}$$

So, the NRGA when t is close to τ, is defined as

$$\Lambda_{nl-RGA}=\Phi.*\Phi^{-T} \tag{6.46}$$

However, the NRGA computation using equation (6.46) is not simple, since it requires the calculation of the Φ matrix. Hence, to propose a practical input-output pairing analysis and to reduce the computational load we can ignore some terms of the matrix Φ. As t is assumed close to τ, these terms include higher order terms ($k>r$), where r is the maximum value of relative orders from y_i to u_j ($i,j=1,\ldots,m$). Using the definition of relative order, the lower order terms ($k<r$) can be ignored as the effect of some inputs on some outputs are not apparent. So, the matrix Φ can be approximated as

$$\Phi\cong\frac{(t-\tau)^r}{r!}\begin{bmatrix} L_{g_1(x)}L_{f(x)}^{r-1}h_1(x) & \cdots & L_{g_m(x)}L_{f(x)}^{r-1}h_1(x)\\ L_{g_1(x)}L_{f(x)}^{r-1}h_2(x) & \cdots & L_{g_m(x)}L_{f(x)}^{r-1}h_2(x)\\ \vdots & & \vdots\\ L_{g_1(x)}L_{f(x)}^{r-1}h_m(x) & \cdots & L_{g_m(x)}L_{f(x)}^{r-1}h_m(x)\end{bmatrix} \tag{6.47}$$

Definition 6.8. The nonlinear RGA is defined as

$$\Lambda_{nl-RGA}=R.*R^{-T} \tag{6.48}$$

where

$$R=\begin{bmatrix} L_{g_1(x)}L_{f(x)}^{r-1}h_1(x) & \cdots & L_{g_m(x)}L_{f(x)}^{r-1}h_1(x)\\ L_{g_1(x)}L_{f(x)}^{r-1}h_2(x) & \cdots & L_{g_m(x)}L_{f(x)}^{r-1}h_2(x)\\ \vdots & & \vdots\\ L_{g_1(x)}L_{f(x)}^{r-1}h_m(x) & \cdots & L_{g_m(x)}L_{f(x)}^{r-1}h_m(x)\end{bmatrix} \tag{6.49}$$

and $r=\max\{r_{ij}\}\le n,\ i,j=1,\ldots,m.$

Note. It is readily seen that the NRGA in equation (6.48) is time independent and has similar properties to the conventional RGA in linear multivariable systems. But Λ_{nl-RGA} is not computed in the steady state and depends on the operating point and dynamical properties of the plant.

6.4.2.1 Pairing Rules

The following rules are proposed to select the appropriate input-output pairs for the affine nonlinear multivariable plants using the NRGA:

- Compute the relative order matrix for all inputs-outputs and determine its largest value r.
- Compute the NRGA using equation (6.48). Select the elements close to 1.0 for the input-output pairing selection.

6.4.3 Linear Interpretation of the Nonlinear-RGA

In section 6.4.2, the NRGA is presented for input-output pairing of affine nonlinear multivariable plants. This matrix can be applied to linear multivariable plants. Suppose that a linear multivariable plant is given by the following state space equations

$$\begin{aligned}
\begin{bmatrix} \dot{x}_1 \\ \vdots \\ \dot{x}_n \end{bmatrix} &= A\begin{bmatrix} x_1 \\ \vdots \\ x_n \end{bmatrix} + \underbrace{\begin{bmatrix} B_{*1} & \cdots & B_{*m} \end{bmatrix}}_{B}\begin{bmatrix} u_1 \\ \vdots \\ u_m \end{bmatrix} \\
\begin{bmatrix} y_1 \\ \vdots \\ y_m \end{bmatrix} &= \begin{bmatrix} C_{1*} \\ \vdots \\ C_{m*} \end{bmatrix} \times \begin{bmatrix} x_1 \\ \vdots \\ x_n \end{bmatrix} = C\begin{bmatrix} x_1 \\ \vdots \\ x_n \end{bmatrix}
\end{aligned} \tag{6.50}$$

where C_{i*} is the ith row of C and B_{*j} is the jth column of B. Comparing equations (6.25) and (6.50), shows that the functions $f(x)$, $g_j(x)$ and $h_i(x)$ for the corresponding linear state space model are

$$\begin{aligned}
f(x) &= Ax \\
g_j(x) &= B_{*j} \\
h_i(x) &= C_{i*}x
\end{aligned} \tag{6.51}$$

Equations (6.49) and (6.51) are employed to compute the matrix R. It can easily be shown that in the linear case, the matrix R is as follows (Moaveni and Khaki-Sedigh 2007)

$$R = \begin{bmatrix} C_{1*}A^{r-1}B \\ C_{2*}A^{r-1}B \\ \vdots \\ C_{n*}A^{r-1}B \end{bmatrix} \tag{6.52}$$

Hence, using equations (6.48) and (6.52), a new approach to the input-output pairing for linear multivariable plants is provided. It is important to observe that the proposed method takes the full dynamic effects of the plant into account. The Λ_{nl-RGA} can be used as a dynamic interaction measure for linear and nonlinear multivariable plants. However, it is important to note that this method is the most effective for nonlinear multivariable plants.

Example 6.4.2

Consider the Quadruple-tank plant introduced in examples 6.3.6 and 6.4.1. We have $r = \max\{r_1, r_2\} = 2,$ so the matrix R is

$$R = \begin{bmatrix} (-\frac{k_1 k_c a_1 \sqrt{g}}{\sqrt{2}A_1^2})\frac{\gamma_1}{\sqrt{h_1}} & (\frac{k_2 k_c a_3 \sqrt{g}}{\sqrt{2}A_1 A_3})\frac{(1-\gamma_2)}{\sqrt{h_3}} \\ (\frac{k_1 k_c a_4 \sqrt{g}}{\sqrt{2}A_2 A_4})\frac{(1-\gamma_1)}{\sqrt{h_4}} & (-\frac{k_2 k_c a_2 \sqrt{g}}{\sqrt{2}A_2^2})\frac{\gamma_2}{\sqrt{h_2}} \end{bmatrix}$$

and the NRGA for $k_1 = k_2$, $A_1 = A_2 = A_3 = A_4$ and $a_1 = a_2 = a_3 = a_4$ is

$$\Lambda_{nl-RGA} = \frac{1}{\frac{\gamma_1\gamma_2}{\sqrt{h_1 h_2}} - \frac{(1-\gamma_1)(1-\gamma_2)}{\sqrt{h_3 h_4}}} \begin{bmatrix} \frac{\gamma_1\gamma_2}{\sqrt{h_1 h_2}} & -\frac{(1-\gamma_1)(1-\gamma_2)}{\sqrt{h_3 h_4}} \\ -\frac{(1-\gamma_1)(1-\gamma_2)}{\sqrt{h_3 h_4}} & \frac{\gamma_1\gamma_2}{\sqrt{h_1 h_2}} \end{bmatrix}$$

This can be rewritten as follows

$$\Lambda_{nl-RGA} = \begin{bmatrix} \dfrac{1}{1-\frac{(1-\gamma_1)(1-\gamma_2)}{\gamma_1\gamma_2}\sqrt{\frac{h_1 h_2}{h_3 h_4}}} & \dfrac{1}{1-\frac{\gamma_1\gamma_2}{(1-\gamma_1)(1-\gamma_2)}\sqrt{\frac{h_3 h_4}{h_1 h_2}}} \\ \dfrac{1}{1-\frac{\gamma_1\gamma_2}{(1-\gamma_1)(1-\gamma_2)}\sqrt{\frac{h_3 h_4}{h_1 h_2}}} & \dfrac{1}{1-\frac{(1-\gamma_1)(1-\gamma_2)}{\gamma_1\gamma_2}\sqrt{\frac{h_1 h_2}{h_3 h_4}}} \end{bmatrix}$$

Let

$$\alpha = \frac{1}{1 - \frac{(1-\gamma_1)(1-\gamma_2)}{\gamma_1\gamma_2}\sqrt{\frac{h_1h_2}{h_3h_4}}}$$

Then

$$\Lambda_{nl-RGA} = \begin{bmatrix} \alpha & 1-\alpha \\ 1-\alpha & \alpha \end{bmatrix}$$

The following two cases are considered:

1. If $\alpha > 1-\alpha$, then the appropriate input-output pair is $(u_1 - y_1, u_2 - y_2)$. This result is easily justified as follows:

$$\alpha > 1-\alpha \Rightarrow \ \alpha > 0.5 \Rightarrow \ \frac{1}{1 - \frac{(1-\gamma_1)(1-\gamma_2)}{\gamma_1\gamma_2}\sqrt{\frac{h_1h_2}{h_3h_4}}} > 0.5 \Rightarrow$$

$$\Rightarrow 1 + \gamma_1\gamma_2(1 - \sqrt{\frac{h_3h_4}{h_1h_2}}) < \gamma_1 + \gamma_2 < \min\{1 + \gamma_1\gamma_2(1 + \sqrt{\frac{h_3h_4}{h_1h_2}}),\ 2\}$$

where it shows that the new result is compatible with the result of linear RGA analysis, $1 < \gamma_1 + \gamma_2 < 2$ in example 2.3.1.

2. If $\alpha < 1-\alpha$, then the appropriate input-output pair is $(u_1 - y_2, u_2 - y_1)$. So, we have

$$\alpha < 1-\alpha \Rightarrow \ \alpha < 0.5 \Rightarrow 0 < \gamma_1 + \gamma_2 < 1 + \gamma_1\gamma_2(1 - \sqrt{\frac{h_3h_4}{h_1h_2}})$$

where the condition is compatible with the linear RGA analysis, $0 < \gamma_1 + \gamma_2 < 1$. It is interesting to note that the above condition shows the dependence of the appropriate input-output pair on the operating point of the nonlinear plant.

To compare the input-output pairing analysis according to the NRGA, with the linear RGA analysis, we consider the physical constants similar to example 6.4.1 but we set the following conditions for the Quadruple-tank process

$$h_1 = 12.7 \ \ (cm), \ \ h_2 = 12.4 \ \ (cm)$$
$$h_3 = 7.4 \ \ (cm), \ \ h_4 = 8.7 \ \ (cm)$$
$$\gamma_1 = 0.48, \ \ \gamma_2 = 0.45$$

For the above conditions, RGA proposes $(u_1 - y_2, u_2 - y_1)$ as an appropriate input-output pair (Johansson 2000). Using the nonlinear RGA, the inequality $0.8782 < \gamma_1 + \gamma_2 = 0.93 < 1.5538$ holds true and $(u_1 - y_1, u_2 - y_2)$ is proposed as the appropriate pair. Similar to example 6.4.1 we use the decentralized control structure with PI controllers introduced by Johansson (2000) to compare the closed loop performances of the proposed input-output pairings. In figure 6.9, responses of the Quadruple-tank according to $(u_1 - y_2, u_2 - y_1)$ and $(u_1 - y_1, u_2 - y_2)$ are shown. Simulation results indicate that the pairing suggested by the NRGA are superior to the off-diagonal pairing proposed by the RGA.

■

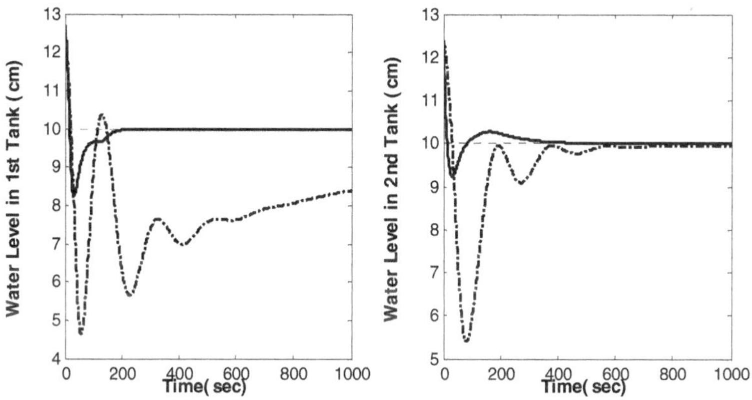

Fig. 6.9 Closed loop step responses of the Quadruple-tank using decentralized control. Solid line corresponds to diagonal pairing; Dash-dot line corresponds to off diagonal pairing; Dashed line is the reference input.

Example 6.4.3

Consider the simplified state space model of a free gyroscope given by the following equations

$$\dot{x}_1 = x_2$$

$$\dot{x}_2 = \frac{1}{I_r \cos x_3}(T_z + I_R \omega_s x_4)$$

$$\dot{x}_3 = x_4$$

$$\dot{x}_4 = \frac{1}{I_r}(T_y - I_R \omega_s x_2 \cos x_3)$$

where $x_1 = \psi_D, x_2 = \dot{\psi}_D, x_3 = \theta_D, x_2 = \dot{\theta}_D$. Plant inputs are the torques $u_1 = T_z$ and $u_2 = T_y$, the outputs are $y_1 = \psi_D$ and $y_2 = \theta_D$, yaw and pitch angels of the rotor respectively. Also, I_r and I_R denote the moment of inertia in pitch and yaw respectively and ω_s shows the rotor angular speed.

The step responses of the plant are shown in figure 6.10 for $I_R = 30\times10^{-6}$, $I_r = 15\times10^{-6}$, $\omega_s = 200\pi$ and $\mathbf{x}(0) = [1.2^o, 0, 1.3^o, 0]$. This shows the instability and oscillations (Nutation) in responses. The corresponding digraph of the gyro is shown in figure 6.11, and its relative-order matrix is

$$M_r = \begin{bmatrix} r_{11} & r_{12} \\ r_{21} & r_{22} \end{bmatrix} = \begin{bmatrix} 2 & 3 \\ 3 & 2 \end{bmatrix}$$

Also, the corresponding characteristic matrix is

$$C(x) = \begin{bmatrix} L_{g_1} L_f h_1(x) & L_{g_2} L_f h_1(x) \\ L_{g_1} L_f h_2(x) & L_{g_2} L_f h_2(x) \end{bmatrix} = \begin{bmatrix} \dfrac{1}{I_r \cos x_3} & 0 \\ 0 & \dfrac{1}{I_r} \end{bmatrix}$$

where it is generically full rank and according to theorem 6.3, $(u_1 - y_1, u_2 - y_2)$ is the appropriate input-output pair. On the other hand, the matrix R is

$$R = \begin{bmatrix} L_{g_1} L_f^2 h_1(x) & L_{g_2} L_f^2 h_1(x) \\ L_{g_1} L_f^2 h_2(x) & L_{g_2} L_f^2 h_2(x) \end{bmatrix} = \begin{bmatrix} 0 & \dfrac{I_R \omega_s}{I_r^2 \cos x_3} \\ -\dfrac{I_R \omega_s}{I_r^2} & 0 \end{bmatrix}$$

and so, the NRGA is

$$\Lambda_{nl-RGA} = \begin{bmatrix} 0 & 1 \\ 1 & 0 \end{bmatrix}$$

where it proposes the $(u_1 - y_2, u_2 - y_1)$ as the appropriate input-output pair. Figure 6.12 shows the closed loop plant step responses using this decentralized control structure and two PID controllers for $(u_1 - y_2, u_2 - y_1)$. The transfer function matrix of the linearized plant around the operating point is given below. It is obvious that the conventional RGA can not be used.

$$G(s) = \begin{bmatrix} \dfrac{8.507\times10^7}{s^3 + 1.579\times10^6 s} & \dfrac{6.77\times10^4}{s^2 + 1.579\times10^6} \\ \dfrac{6.667\times10^4}{s^2 + 1.579\times10^6} & -\dfrac{8.378\times10^7}{s^3 + 1.579\times10^6 s} \end{bmatrix}$$

■

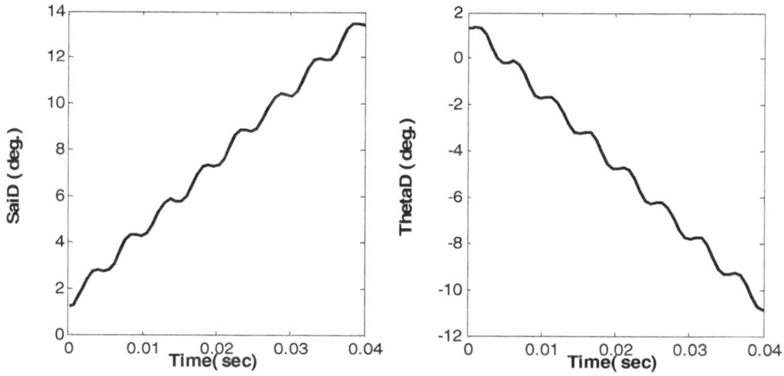

Fig. 6.10 Step responses of the free gyroscope.

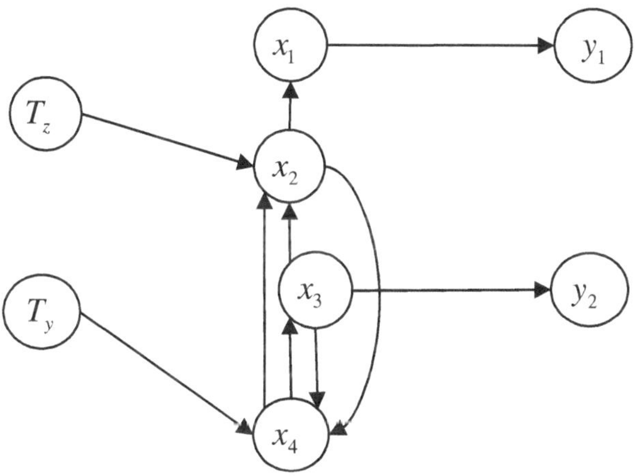

Fig. 6.11 The Digraph of free gyroscope.

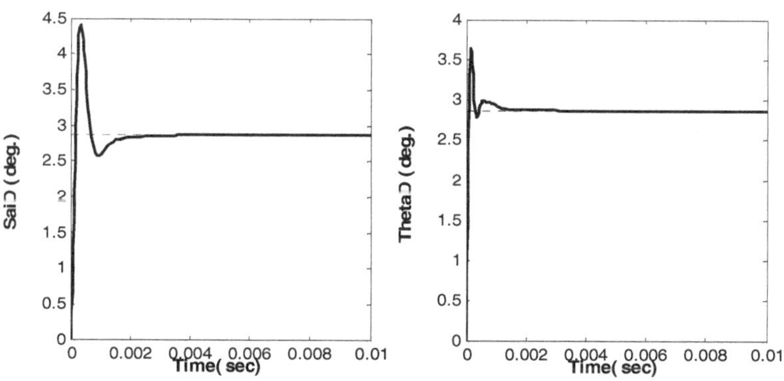

Fig. 6.12 Closed loop step responses corresponding to off-diagonal pairing for decentralized control.

Example 6.4.4

Consider the following distillation column transfer function matrix model (Wood and Berry 1973)

$$\begin{bmatrix} X_D(s) \\ X_B(s) \end{bmatrix} = \begin{bmatrix} \dfrac{12.8e^{-s}}{16.7s+1} & \dfrac{-18.9e^{-s}}{21s+1} \\ \dfrac{6.6e^{-7s}}{10.9s+1} & \dfrac{-19.4e^{-s}}{14.4s+1} \end{bmatrix} \begin{bmatrix} R(s) \\ S(s) \end{bmatrix}$$

where X_D and X_B are the overhead and bottoms compositions of methanol respectively. Also, R is the reflux flow rate and S is the steam flow rate to the reboiler.

A state space realization of the transfer function matrix using a 1st order Pade approximation for the delay is

$$A = \begin{bmatrix} -2.4373 & -0.9235 & -0.0992 & 0.0024 & -0.0007 & 0.0011 & -0.0016 \\ 1 & 0 & 0 & 0 & 0 & 0 & 0 \\ 0 & 1 & 0 & 0 & 0 & 0 & 0 \\ 0 & 0 & -0.7603 & -0.0003 & -0.5054 & 0.2820 & 0.4122 \\ 0 & 0 & 0.2253 & -0.0014 & -1.3007 & -0.6874 & 0.1177 \\ 0 & 0 & -0.3379 & -0.0001 & 0.7754 & 0.1253 & -0.2612 \\ 0 & 0 & 0.5069 & 0.0002 & 0.3369 & 0.8120 & 0.3919 \end{bmatrix}$$

$$B = \begin{bmatrix} 1 & 0 \\ 0 & 0 \\ 0 & 0 \\ 0 & 0.1713 \\ 0 & 0.9492 \\ 0 & 0.0761 \\ 0 & -0.1142 \end{bmatrix}$$

$$C = \begin{bmatrix} -0.7665 & 1.2436 & 0.5585 & -0.0583 & 0.9091 & 0.0489 & -0.3796 \\ -0.6055 & -1.0743 & 0.2838 & -0.0443 & 1.3519 & 0.0572 & -0.5883 \end{bmatrix}$$

The relevant relative-order matrix is

$$M = \begin{bmatrix} r_{11} & r_{12} \\ r_{21} & r_{22} \end{bmatrix} = \begin{bmatrix} 1 & 1 \\ 1 & 1 \end{bmatrix}$$

And the relative-order matrix can not propose any control configuration. However, in (Chen and Seborg 2002) it is shown that the corresponding RGA is

$$\Lambda = \begin{bmatrix} 2.01 & -1.01 \\ -1.01 & 2.01 \end{bmatrix}$$

and $(u_1 - y_1, u_2 - y_2)$ is the appropriate input-output pair.

Next, we calculate the NRGA. This gives

$$\Lambda_{nl-RGA} = \begin{bmatrix} 2.1175 & -1.1175 \\ -1.1175 & 2.1175 \end{bmatrix}$$

where it proposes the same pairing as the RGA.

■

6.5 Conclusion

In this chapter, control configuration techniques for nonlinear affine multivariable plants have been developed. Some key definitions and results in the nonlinear system theory were given for later use in the subsequent chapters. A digraph representation was introduced with the relative order matrix notion to present a structural interaction measure. This led to a pairing methodology for the nonlinear plants.

Also, in other two different approaches, first the RGA tool for linear plants was extended to the nonlinear plants and then the NRGA was defined. Several worked examples and simulation results were presented. It was shown that the relative order matrix and the NRGA are the most effective for pairing selection in affine nonlinear multivariable plants. It was also shown that pairing in nonlinear plants can change with the operating point of the plant.

References

Chen, D., Seborg, D.E.: Relative gain array analysis for uncertain process models. AIChE J. 48, 302–310 (2002)

Daoutidis, P., Kravaris, C.: Structural evaluation of control configuration for multivariable nonlinear processes. Chem. Eng. Sci. 47, 1091–1107 (1992)

Gelb, A., Vander Velde, W.E.: Multiple-Input describing functions and nonlinear system design. McGraw-Hill, New York (1968)

Glad, S.T.: Extensions of the RGA concept to nonlinear systems. In: Proceeding of the 5th European Control Conference, Karlsruhe, Germany (1999)

Isidori, A.: Nonlinear control systems. Springer, New York (1995)

Johansson, K.H.: The Quadruple-tank process: A multivariable laboratory process with an adjustable zero. IEEE Trans. Control Syst. Technol. 8, 456–465 (2000)

Moaveni, B., Khaki-Sedigh, A.: Input-output pairing for nonlinear multivariable systems. J. Applied Sciences 7, 3492–3498 (2007)

Nijmeijer, H., Van de Schaft, A.: Nonlinear dynamical control systems. Springer, New York (1996)

Slotine, J.-J.E., Li, W.: Applied nonlinear control. Prentice Hall, New Jersey (1991)

Wood, R.K., Berry, M.W.: Terminal composition control of a binary distillation column. Chem. Eng. Sci. 28, 1707–1717 (1973)

Chapter 7
Control Configuration Selection of Linear Uncertain Multivariable Plants

7.1 Introduction

Modeling of complex multivariable plants is always subject to uncertainty in their linear models. However, in the face of unknown or uncertain multivariable plants, the control configuration of the plant may endure fundamental changes, which will severely degrade the decentralized controller performance. The well-known input-output pairing techniques described in previous sections are unable to analyze the effect of uncertainty on input-output pairing and only recently, pairing methods are proposed for uncertain multivariable plants. The approaches to control configuration selection in the presence of uncertainty can be categorized in the following classes:

- Structured uncertainties.
- Diagonal input uncertainties.
- Condition number and robustness analysis.
- Statistical description of uncertainty bound for the RGA.
- Unstructured uncertainty modeling and bounds on the magnitude of the worst-case relative gain.
- State space description of uncertain multivariable plants and the DIOPM approach.
- On-line identification and adaptive input-output pairing to encounter parameter changes.

This chapter is organized as follows. In section 7.2, a motivating example is used to show the problems associated with the input-output pairing of uncertain multivariable plants. Then, the basic results available on structured uncertainty, input uncertainty and condition number in the multivariable transfer function are presented. Section 7.3 provides the statistical based approach to uncertainty and control configuration selection, which is the first serious result in the input-output pairing of uncertain multivariable plants. Pairing in norm-bounded uncertain plants is studied in section 7.4. The uncertain multivariable plant is modeled with unstructured additive norm-bounded uncertainties, and a bound on the magnitude

A. Khaki-Sedigh and B. Moaveni: Control Configuration Selection, LNCIS 391, pp. 173–217.
springerlink.com

of the worst-case relative gain, calculated at steady state and also at higher frequencies is derived. Section 7.5 deals with state space uncertain models. Based on these time domain models, pairing of uncertain multivariable plants using the DIOPM is presented. Finally, adaptive input-output pairing and a simple reconfigurable decentralized controller is developed in section 7.6.

7.2 The Preliminary Approaches

In this section, we first present a motivating example to show the effect of uncertainty or parameter variation on the RGA.

Example 7.2.1

Consider the following transfer function matrix (Haggblom 1995)

$$G(s) = \frac{1}{10s+1}\begin{bmatrix} 1 & 10 \\ 0 & 1 \end{bmatrix}$$

and it is obvious that the corresponding RGA is the identity matrix. A small change in the (2,1) entry of $G(s)$ is considered

$$G_{P1}(s) = \frac{1}{10s+1}\begin{bmatrix} 1 & 10 \\ 0.09 & 1 \end{bmatrix}$$

and the corresponding RGA becomes

$$\Lambda_{P1} = \begin{bmatrix} 10 & -9 \\ -9 & 10 \end{bmatrix}$$

Diagonal pairing is the appropriate input-output pair for $G(s)$ and $G_{P1}(s)$. However, the closed loop step responses of these transfer functions using an identical PI controller are shown in figure 7.1. It is obvious that responses of $G_{P1}(s)$ are more sluggish than $G(s)$.

It is interesting to note that if the element (2,1) of $G(s)$ changes to 0.2

$$G_{P2}(s) = \frac{1}{10s+1}\begin{bmatrix} 1 & 10 \\ 0.2 & 1 \end{bmatrix}$$

the corresponding RGA becomes

$$\Lambda_{P2} = \begin{bmatrix} -1 & 2 \\ 2 & -1 \end{bmatrix}$$

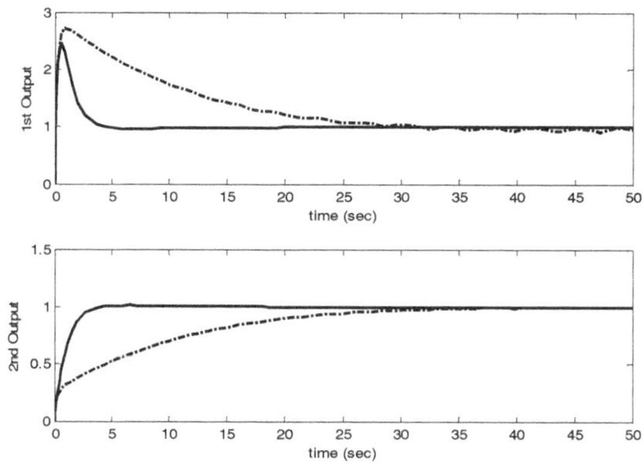

Fig. 7.1 The closed loop step responses of $G(s)$ and $G_{P1}(s)$. Solid lines and Dash-dot lines correspond to $G(s)$ and $G_{P1}(s)$, respectively.

It is obvious that off-diagonal pairing is now the appropriate input-output pair. These results show that input-output pairing is sensitive to modeling errors or uncertainty, and the plant may become fundamentally more difficult to control.

■

7.2.1 Structured Uncertainties and RGA Robustness

Parametric or structured uncertainty is in some cases effectively used to model the uncertain plants. In this section, control configuration selection of uncertain multivariable plants with parametric uncertainty is considered.

In the case of independent element uncertainty in the multivariable transfer function matrix $G(s)$, a sufficient and necessary condition for the singularity of the transfer function matrix was first proved in (Yu and Luyben 1987) and later in (Hovd and Skogestad 1992). This is presented in section 2.2.2 as the 7th RGA property, and is restated in the next theorem.

Theorem 7.1

The transfer function matrix $G(s)$ becomes singular if a single element in $G(s)$ is perturbed from $g_{ij}(s)$ to $g_{pij}(s) = g_{ij}(s)(1-\frac{1}{\lambda_{ij}})$. That is equivalent to a relative change $-\frac{1}{\lambda_{ij}}$ in the $(i,j)^{th}$ element of $G(s)$.

■

Example 7.2.2

Consider a 3×3 linear multivariable plant with the following transfer function matrix

$$G(s) = \begin{bmatrix} \frac{119e^{-5s}}{21.7s+1} & \frac{40e^{-5s}}{337s+1} & \frac{-2.1e^{-5s}}{10s+1} \\ \frac{77e^{-5s}}{50s+1} & \frac{76.7e^{-3s}}{28s+1} & \frac{-5e^{-5s}}{10s+1} \\ \frac{93e^{-5s}}{50s+1} & \frac{-36.7e^{-5s}}{166s+1} & \frac{-103.3e^{-4s}}{23s+1} \end{bmatrix}$$

and the corresponding RGA is

$$\Lambda = \begin{bmatrix} 1.4975 & -0.4650 & -0.0325 \\ -0.5031 & 1.4403 & 0.0628 \\ 0.0056 & 0.0247 & 0.9697 \end{bmatrix}$$

If the element (2,2) of $G(s)$ is perturbed as

$$G_p(s) = \begin{bmatrix} \frac{119e^{-5s}}{21.7s+1} & \frac{40e^{-5s}}{337s+1} & \frac{-2.1e^{-5s}}{10s+1} \\ \frac{77e^{-5s}}{50s+1} & \frac{76.7e^{-3s}}{28s+1}\boxed{(1-\frac{1}{1.4403})} & \frac{-5e^{-5s}}{10s+1} \\ \frac{93e^{-5s}}{50s+1} & \frac{-36.7e^{-5s}}{166s+1} & \frac{-103.3e^{-4s}}{23s+1} \end{bmatrix}$$

then $\det(G_p(0)) = 0$ and $G_p(0)$ will be singular.

■

Note 1. As the elements of physical plants are always coupled, the assumption of individual element uncertainty does not hold in many practical plants.

Note 2. The plants with larger relative gains are more sensitive to modeling uncertainty.

In the face of different parametric uncertainties in the plant, each transfer function element is modeled by uncertain parameters. The following four examples derived from (Khaki-Sedigh and Moaveni 2003a), show the control configuration selection problem in such cases for 2×2 and 3×3 transfer function matrices.

Example 7.2.3

Consider a 2×2 multivariable plant $G(s)$. The (1,1) entry of the RGA element can be written as

$$\lambda_{11} = \frac{g_{11}g_{22}}{g_{11}g_{22} - g_{12}g_{21}} \tag{7.1}$$

Defining $k = \frac{g_{12}g_{21}}{g_{11}g_{22}}$, gives

$$\lambda_{11} = \frac{1}{1-k} \tag{7.2}$$

and similarly λ_{12} can be written as

$$\lambda_{12} = \frac{-k}{1-k} \tag{7.3}$$

It is clear that the input-output pairing can be determined by comparing the elements of the first row of the RGA. Hence, if $\lambda_{11} > \lambda_{12}$ it follows that

$$-1 < k < 1 \tag{7.4}$$

In the face of parametric uncertainties, k will be an uncertain parameter and can be presented as

$$k' = k + \Delta k \tag{7.5}$$

If Δk causes a change in the previous input-output pairing, i.e. $\lambda_{12} > \lambda_{11}$ then

$$\frac{1}{1-k'} < \frac{-k'}{1-k'} \tag{7.6}$$

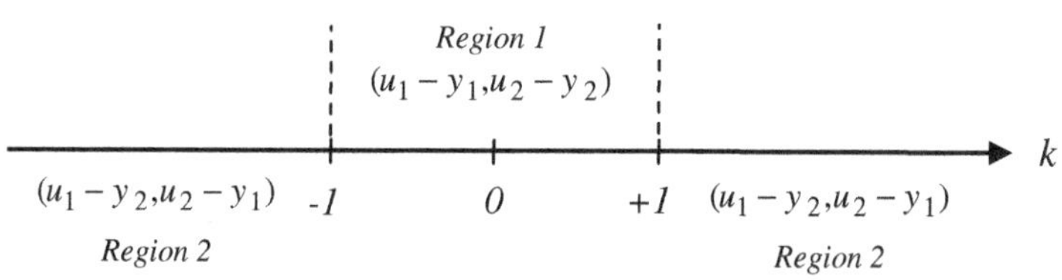

Fig. 7.2 Variation bound of k and its affect on input-output pairing for 2×2 uncertain plants.

which gives $k' > 1$ or $k' < -1$, and this is shown graphically in figure 7.2. A transfer from region 1 to 2 or vice versa implies a change in the input-output pairing due to the parametric uncertainty.

■

Example 7.2.4

Consider the following transfer function matrix

$$G(s) = \begin{bmatrix} \dfrac{(12.8+\delta_{11})e^{-s}}{16.7s+1} & \dfrac{-(18.9+\delta_{12})e^{-3s}}{21s+1} \\ \dfrac{(6.6+\delta_{21})e^{-7s}}{10.9s+1} & \dfrac{-(19.4+\delta_{22})e^{-3s}}{14.4s+1} \end{bmatrix}$$

where $\delta_{11} \in [-6,2]$, $\delta_{12} \in [-7,3]$, $\delta_{21} \in [-1,3]$, $\delta_{22} \in [-2,2]$ and its corresponding RGA for the nominal case $\delta_{ij} = 0$ $(i, j = 1,2)$ is

$$\Lambda_1 = \begin{bmatrix} 2.01 & -1.01 \\ -1.01 & 2.01 \end{bmatrix}$$

It follows from example 7.2.3 that $k = 0.5023$. The RGA matrix shows that $(u_1 - y_1, u_2 - y_2)$ is an appropriate input-output pair. For $\delta_{11} = -4$, $\delta_{12} = -5$, $\delta_{21} = 1$, $\delta_{22} = 0$, parameter k in example 7.2.3 is changed to

$$k' = \frac{(6.6+1)(-18.9-5)}{(12.8-4)(-19.4)} = 1.064 > 1$$

which clearly indicates that a change in the input-output pairing has occurred, as shown in figure 7.2. This result is also verified by calculating the RGA matrix for the new parameters, i.e.

$$\Lambda_2 = \begin{bmatrix} -15.63 & 16.63 \\ 16.63 & -15.63 \end{bmatrix}$$

which also shows a change in the input-output pairing.

■

Example 7.2.5

Consider a 3×3 multivariable plant $G(s)$. The ith row of the RGA elements can be written as

$$\lambda_{i1} = \frac{g_{i1}\det(G^{i1})}{g_{i1}\det(G^{i1}) - g_{i2}\det(G^{i2}) + g_{i3}\det(G^{i3})}$$

$$\lambda_{i2} = \frac{-g_{i2}\det(G^{i2})}{g_{i1}\det(G^{i1}) - g_{i2}\det(G^{i2}) + g_{i3}\det(G^{i3})}$$

$$\lambda_{i3} = \frac{g_{i3}\det(G^{i3})}{g_{i1}\det(G^{i1}) - g_{i2}\det(G^{i2}) + g_{i3}\det(G^{i3})}$$

and by defining $k_{i1} = \dfrac{g_{i2}\det(G^{i2})}{g_{i1}\det(G^{i1})}$ and $k_{i2} = \dfrac{g_{i3}\det(G^{i3})}{g_{i1}\det(G^{i1})}$, we have

$$\lambda_{i1} = \frac{1}{1 + k_{i2} - k_{i1}}$$

$$\lambda_{i2} = \frac{-k_{i1}}{1 + k_{i2} - k_{i1}}$$

$$\lambda_{i3} = \frac{k_{i2}}{1 + k_{i2} - k_{i1}}$$

The regions in the $k_1 - k_2$ coordinates ($k_1 = k_{i1},\ k_2 = k_{i2},\ i = 1,2,3$) which indicate a change in the input-output pairings are now determined.

Case 1. ($\lambda_{i1} > \lambda_{i2}, \lambda_{i1} > \lambda_{i3}$)

In this case, a closed region is characterized by

$$1 + k_{i2} - k_{i1} > 0$$
$$k_{i1} > -1$$
$$k_{i2} < 1$$

and is shown in figure 7.3. The shaded region in figure 7.3 represents the inequalities and shows that $(u_1 - y_i)$ is an appropriate pair, but a final decision must be made after considering the other cases.

Case 2. ($\lambda_{i2} > \lambda_{i1}, \lambda_{i2} > \lambda_{i3}$)

In this case, we have

$$\begin{array}{ll} 1 + k_{i2} - k_{i1} > 0 & 1 + k_{i2} - k_{i1} < 0 \\ k_{i1} < -1 & ,\ k_{i1} > -1 \\ k_{i2} + k_{i1} < 0 & k_{i2} + k_{i1} > 0 \end{array}$$

and is shown in figure 7.4.

Case 3 ($\lambda_{i3} > \lambda_{i1}, \lambda_{i3} > \lambda_{i2}$)

In this case, we have

$$\begin{array}{ll} 1+k_{i2}-k_{i1}>0 & 1+k_{i2}-k_{i1}<0 \\ k_{i2}>1 & ,\ k_{i2}<1 \\ k_{i2}+k_{i1}>0 & k_{i2}+k_{i1}<0 \end{array}$$

and this region is shown in figure 7.5.

■

Based on the above observations, a test procedure is derived.

1. Determine the $\hat{k}_i = (k_{i1}, k_{i2})$ variables.
2. Identify the points $\hat{k}_1$, $\hat{k}_2$ and $\hat{k}_3$ in the $k_1 - k_2$ diagram.
3. A shift of $\hat{k}_1$, $\hat{k}_2$ and $\hat{k}_3$ from one region to another may indicate a change in the input-output pairing.

Note. If any two of the $\hat{k}_1$, $\hat{k}_2$ and $\hat{k}_3$ or all three lie in the same region, then a block decentralized control structure may be preferred.

Example 7.2.6

Consider the following 3×3 transfer function matrix

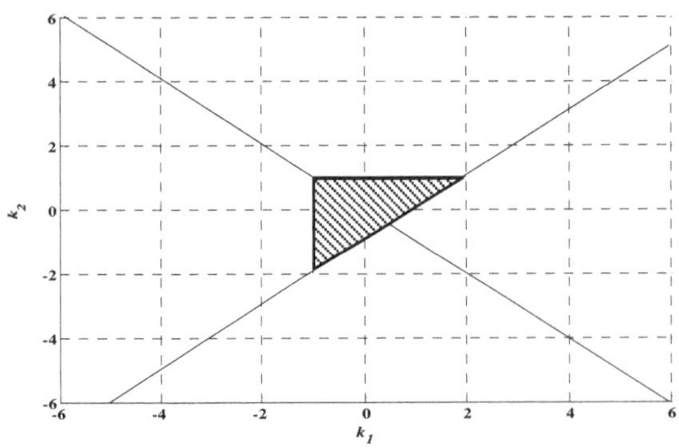

Fig. 7.3 The region in the $k_1 - k_2$ diagram corresponding to $\lambda_{i1} > \lambda_{i2}, \lambda_{i1} > \lambda_{i3}$ for 3×3 plants.

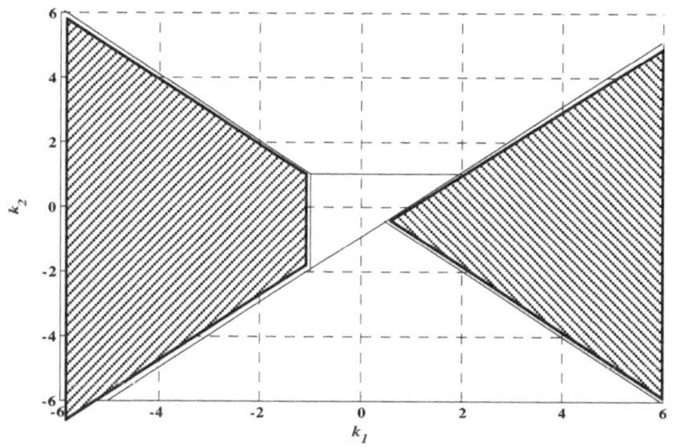

Fig. 7.4 The region in the $k_1 - k_2$ diagram corresponding to $\lambda_{i2} > \lambda_{i1}, \lambda_{i2} > \lambda_{i3}$ for 3×3 plants.

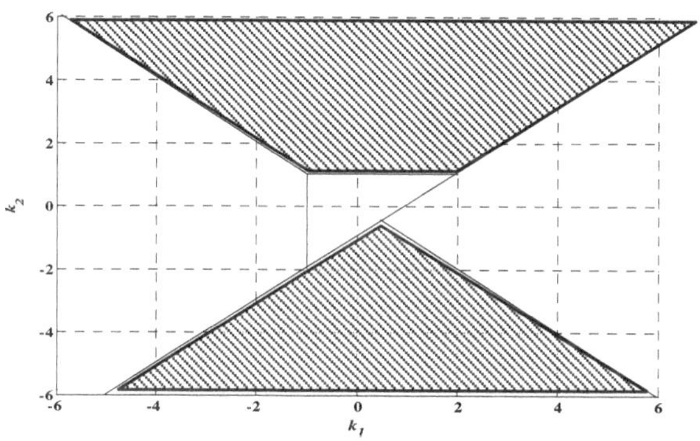

Fig. 7.5 The region in the $k_1 - k_2$ diagram corresponding to $\lambda_{i3} > \lambda_{i1}, \lambda_{i3} > \lambda_{i2}$ for 3×3 plants.

$$G(s) = \begin{bmatrix} \dfrac{0.66e^{-2.6s}}{6.7s+1} & \dfrac{-0.61e^{-3.5s}}{8.64s+1} & \dfrac{-0.0049e^{-s}}{9.06s+1} \\ \dfrac{1.11e^{-0.65s}}{3.25s+1} & \dfrac{-2.36e^{-3s}}{5s+1} & \dfrac{-0.012e^{-1.2s}}{7.09s+1} \\ \dfrac{-33.68e^{-9.2s}}{8.15s+1} & \dfrac{46.2e^{-9.4s}}{10.9s+1} & \dfrac{0.87(11.61s+1)e^{-s}}{(3.89s+1)(18.8s+1)} \end{bmatrix}$$

Figure 7.6 represents the positions of $\hat{k}_i$, $i = 1, 2, 3$ for the nominal plant, and its corresponding RGA is

$$\Lambda_1 = \begin{bmatrix} 1.945 & -0.674 & -0.272 \\ -0.664 & 1.899 & -0.235 \\ -0.281 & -0.225 & 1.506 \end{bmatrix}$$

Case 1. Let the dc gain of the nominal plant be changed as

$$G_2(0) = \begin{bmatrix} 0.95 & 1.05 & 0.95 \\ 1.05 & 0.95 & 0.95 \\ 1.05 & 0.95 & 1.05 \end{bmatrix} .*G(0)$$

Figure 7.7 shows the positions of $\hat{k}_1$, $\hat{k}_2$ and $\hat{k}_3$ which indicates that a change in the input-output pairing has not occurred. This is verified by its RGA matrix

$$\Lambda_2 = \begin{bmatrix} 2.3341 & -1.0191 & -0.3150 \\ -1.0675 & 2.2010 & -0.1335 \\ -0.2666 & -0.1818 & 1.4485 \end{bmatrix}$$

Case 2. Let the dc gain of the nominal plant be also changed as follows

$$G_3(0) = \begin{bmatrix} 0.6 & 1.2 & 0.8 \\ 1.2 & 0.9 & 0.8 \\ 1.2 & 0.8 & 1.2 \end{bmatrix} .*G(0)$$

Figure 7.8 shows the positions of $\hat{k}_1$, $\hat{k}_2$ and $\hat{k}_3$ which indicates that a change has occurred in the input-output pairing.

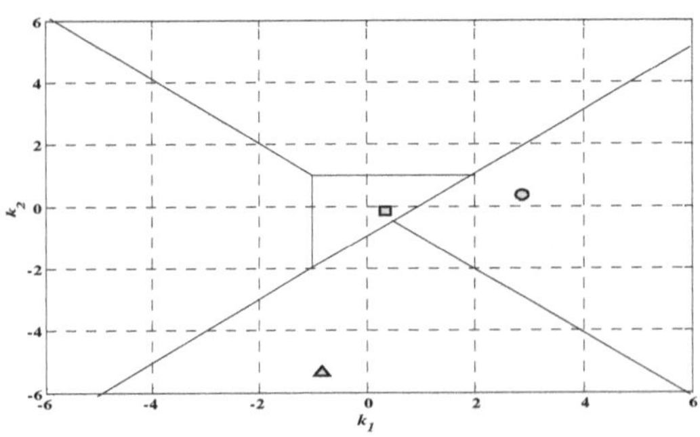

Fig. 7.6 The positions of $\hat{k}_i$, $i = 1,2,3$ in the $k_1 - k_2$ diagram corresponding to the nominal plant. The square symbol denotes $\hat{k}_1$, circle denotes $\hat{k}_2$ and triangle denotes $\hat{k}_3$.

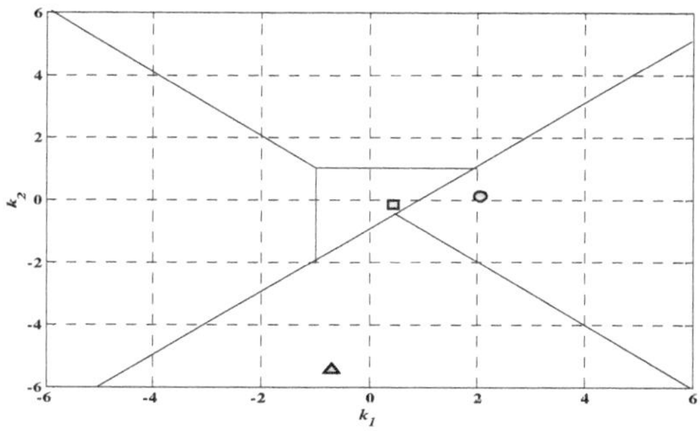

Fig. 7.7 The positions of $\hat{k}_i$, $i = 1,2,3$ in the $k_1 - k_2$ diagram corresponding to the perturbed plant. The square symbol denotes $\hat{k}_1$, circle denotes $\hat{k}_2$ and triangle denotes $\hat{k}_3$.

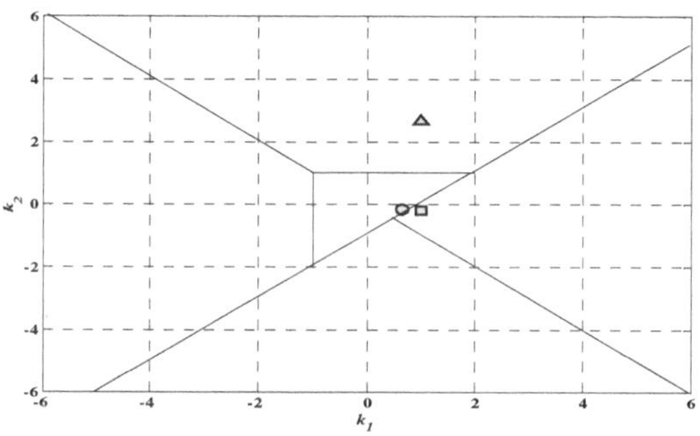

Fig. 7.8 The positions of $\hat{k}_i$, $i = 1,2,3$ in the $k_1 - k_2$ diagram corresponding to the perturbed plant. The square symbol denotes $\hat{k}_1$, circle denotes $\hat{k}_2$ and triangle denotes $\hat{k}_3$.

rresponding RGA matrix given byThis is also verified by the co

$$\Lambda_3 = \begin{bmatrix} -5.275 & 5.249 & 1.026 \\ 5.900 & -3.873 & -1.026 \\ 0.375 & -0.375 & 1.00 \end{bmatrix}$$

■

Note. The results presented in the above examples are limited to 2×2 and 3×3 multivariable plants. Extensions to multivariable plants with more inputs and outputs are not available.

7.2.2 Input Uncertainties and RGA Robustness

Input uncertainty arises from uncertainties associated with the actuators. This is an important form of uncertainties since it is nearly always present in physical plants. Let $\bar{u}_i$ denote the value of the ith input determined by the controller. Then the actual input of the plant will be $u_i = \bar{u}_i(1+\Delta_i)$, where Δ_i represents the respective uncertainty in the ith input channel. Assuming no coupling between the manipulated inputs and defining $\Delta = diag\{\Delta_i\}$, the input uncertain model for the transfer function matrix $G(s)$ is

$$G_p(s) = G(s)(I + \Delta) \tag{7.7}$$

Consider the closed loop control configuration shown in figure 7.9, where $C(s)$ is the controller transfer function matrix, R is the reference input and D is the output or load disturbance. Then, we have

$$Y(s) = \left(I + G_p(s)C(s)\right)^{-1} D(s) \tag{7.8}$$

where

$$G_p(s)C(s) = G(s)(I + \Delta)C(s) = G(s)C(s) + G(s)\Delta C(s) \tag{7.9}$$

Equation (7.9) can be rewritten in one of the following forms

$$G_p(s)C(s) = G(s)C(s)\left(I + C^{-1}(s)\Delta C(s)\right) \tag{7.10}$$

or

$$G_p(s)C(s) = \left(I + G(s)\Delta G^{-1}(s)\right)G(s)C(s) \tag{7.11}$$

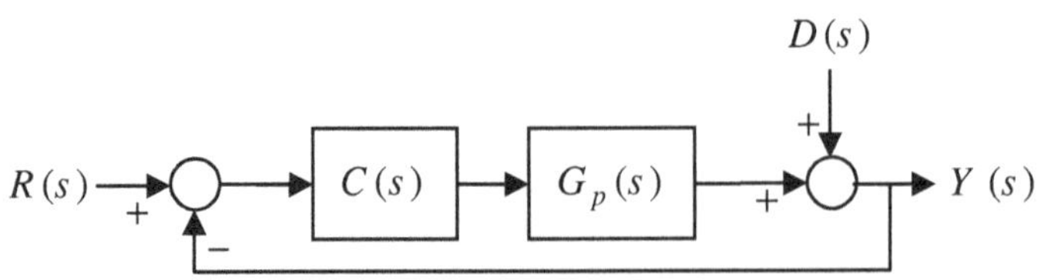

Fig. 7.9 Closed loop control configuration.

Note that in equations (7.10) and (7.11), the actual loop gain is written in terms of the nominal loop gain and an error term. The following result for 2×2 plants is introduced in (Skogestad and Morari 1987). The error term in equation (7.10) and (7.11), can be expressed in terms of the RGA of the controller transfer function matrix as

$$C^{-1}(s)\Delta C(s)=\begin{bmatrix}\lambda_{11}(C(s))\Delta_1+\lambda_{21}(C(s))\Delta_2 & \lambda_{11}(C(s))\dfrac{c_{12}(s)}{c_{11}(s)}(\Delta_1-\Delta_2)\\ -\lambda_{11}(C(s))\dfrac{c_{21}(s)}{c_{22}(s)}(\Delta_1-\Delta_2) & \lambda_{12}(C(s))\Delta_1+\lambda_{11}(C(s))\Delta_2\end{bmatrix} \quad (7.12)$$

where $\lambda_{ij}(C(s))$ is the $(i,j)^{th}$ element of the controller RGA, and

$$G(s)\Delta G^{-1}(s)=\begin{bmatrix}\lambda_{11}(G(s))\Delta_1+\lambda_{12}(G(s))\Delta_2 & -\lambda_{11}(G(s))\dfrac{g_{12}(s)}{g_{22}(s)}(\Delta_1-\Delta_2)\\ \lambda_{11}(G(s))\dfrac{g_{21}(s)}{g_{22}(s)}(\Delta_1-\Delta_2) & \lambda_{12}(G(s))\Delta_1+\lambda_{11}(G(s))\Delta_2\end{bmatrix} \quad (7.13)$$

where $\lambda_{ij}(G(s))$ is the $(i,j)^{th}$ element of the plant RGA.

Note 1. If the controller has small RGA elements, the elements in the error term $C^{-1}(s)\Delta C(s)$ are similar to Δ in magnitude. Hence, $G_p(s)C(s)$ is not particularly influenced by input uncertainty, regardless of how large are the size of the plant RGA elements.

Note 2. If the plant has small RGA elements, the elements in the error term $G(s)\Delta G^{-1}(s)$ are similar to Δ in magnitude. Hence, $G_p(s)C(s)$ is not particularly influenced by input uncertainty, regardless of the size of how large are the controller RGA elements.

Note 3. For a plant to be sensitive to input uncertainty, both the controller and plant RGA elements must be large.

Note 4. Some multivariable control strategies employ an inverse of the plant transfer function matrix to first decouple and then effectively control the SISO loops. That is, $C(s)=G^{-1}(s)K(s)$ where $K(s)$ is a diagonal controller. In this case, $\Lambda(C(s))=\Lambda(G^{-1}(s)K(s))$ and since the controller is diagonal we have $\Lambda(C(s))=\Lambda(G^{-1}(s))=\Lambda^T(G(s))$. Hence, large $\Lambda(G(s))$ will lead to large

$\Lambda(C(s))$ and this indicates high sensitivity to input uncertainty. This partially explains the fact regarding robustness problems with decoupling control strategies.

Example 7.2.7

Consider a distillation column plant presented by the following transfer function matrix (Ogunnaike and Pay 1994)

$$G(s) = \begin{bmatrix} \dfrac{-21.6e^{-s}}{8.5s+1} & \dfrac{1.26e^{-0.3s}}{7.05s+1} \\ \dfrac{-2.75e^{-1.8s}}{8.2s+1} & \dfrac{-4.28e^{-0.35s}}{9.0s+1} \end{bmatrix}$$

The RGA is

$$\Lambda = \begin{bmatrix} 0.964 & 0.036 \\ 0.036 & 0.964 \end{bmatrix}$$

Assume that the relative input uncertainty is only presented in the 2nd input as $\Delta_1 = 0$ and $\Delta_2 = \delta$. Hence, the $G(0)\Delta G^{-1}(0)$ is

$$G(0)\Delta G^{-1}(0) = \begin{bmatrix} 0.036 & -0.284 \\ -0.123 & 0.964 \end{bmatrix} \delta$$

and it is obvious that the magnitude of the elements are less than or equal to δ due to small relative gains.

■

7.2.3 *Condition Number and the Uncertainty Analysis*

An alternative to the RGA for interaction analysis in multivariable plants is the condition number. Condition number is a valuable characteristic of the transfer function matrix $G(s)$ and is defined as

$$\text{cond}(G) \triangleq \frac{\bar{\sigma}(G)}{\underline{\sigma}(G)} \tag{7.14}$$

where $\bar{\sigma}(G)$ and $\underline{\sigma}(G)$ represent the maximum and minimum singular values of G, respectively. Plant condition number is frequency dependent and it is used to identify potential control problems in controlling the multivariable plants.

Some of the condition number properties and implications are now summarized without proof. For more details and proofs, refer to (Skogestad and Postlethwaite 2005).

- In the case of square nonsingular matrices, the condition number measures the difficulty of matrix inversion.
- As $\underline{\sigma}(G) = \dfrac{1}{\overline{\sigma}(G^{-1})}$ it can be easily shown that the condition number is large if both G and G^{-1} have large elements.
- Contrary to the important scaling independence property of the RGA, the condition number is input-output scaling dependant. That is, the condition number of $G_s = D_1 G D_2$, where $D_1 = diag\{d_{1i}\}$ and $D_2 = diag\{d_{2i}\}$ ($i = 1, \cdots, m$) are diagonal matrices, and G are in general different. Hence, the minimized or optimal condition number is defined as

$$\text{cond}^*(G) = \min_{D_1, D_2} (\text{cond}(G_s)) \tag{7.15}$$

A procedure for calculating $\text{cond}^*(G)$ is given in (Skogestad and Postlethwaite 2005).
- For 2×2 multivariable plants, the minimized condition number can be calculated as

$$\text{cond}^*(G) = \|\Lambda\|_{i1} + \sqrt{\|\Lambda\|_{i1}^2 - 1} \tag{7.16}$$

where $\|\Lambda\|_{i1}$ is the induced 1-norm (maximum column sum) of the plant RGA (Grosdidier et al. 1985).
- The sum-norm of the RGA, $\|\Lambda\|_1 = \sum_{i,j} |\lambda_{ij}|$ is approximately the same as the optimal condition number. It is commonly conjectured that $\text{cond}^*(G)$ and the magnitude of the RGA elements are closely related. This is illustrated by the following inequalities (Hovd and Skogestad 1992)

$$\|\Lambda\|_m - \frac{1}{\text{cond}^*(G)} \le \text{cond}^*(G) \le \|\Lambda\|_1 + k(m) \tag{7.17}$$

where $\|\Lambda\|_m \triangleq 2 \max\{\|\Lambda\|_{i1}, \|\Lambda\|_{i\infty}\}$ and $k(m)$ is a constant and $\|\bullet\|_{i\infty}$ denotes the induced infinity-norm (maximum row sum). The lower bound is proved in (Nett and Manousiouthakis 1987) and the upper bound is proved for $m = 2$ with $k(2) = 0$ (Grosdidier et al. 1985). For the general cases $k(3) = 1$ and

$k(4)=2$ are conjectured in (Skogestad and Morari 1987) and (Nett and Manousiouthakis 1987).

- A plant with large condition number is ill conditioned and plants with large RGA elements are therefore always ill conditioned. However, a large condition number may correspond to small RGA elements.
- Small condition number around the crossover frequency is indicative of robust performance in the face of both diagonal and full block input uncertainty.
- Small optimal condition number around the crossover frequency is indicative of robust performance in the face of diagonal input uncertainty.

Example 7.2.8

Consider the dc gain of a 2×2 multivariable plant

$$G(0)=\begin{bmatrix} 40 & 1 \\ 0.1 & 2 \end{bmatrix}$$

and its RGA is

$$\Lambda=\begin{bmatrix} 1.001 & -0.001 \\ -0.001 & 1.001 \end{bmatrix}$$

where the RGA is approximately the identity matrix and has no large elements. But, the condition number of $G(0)$ is large

$$\mathrm{cond}(G(0))=\frac{\bar{\sigma}(G(0))}{\underline{\sigma}(G(0))}=20$$

This means that the plant is sensitive to full-block input uncertainty. Using (7.16) the minimized condition number is $\mathrm{cond}^*(G(0))=1.07$.

■

Example 7.2.9

Consider the dc gain of a 3×3 multivariable plant

$$G(0)=\begin{bmatrix} 1 & 0.9 & 1 \\ 2 & 1 & 2 \\ 0.9 & 0.2 & 1.1 \end{bmatrix}$$

and the corresponding RGA as

$$\Lambda = \begin{bmatrix} -4.375 & 2.25 & 3.125 \\ 9.875 & -1.25 & -7.625 \\ -4.5 & 0 & 5.5 \end{bmatrix}$$

It is obvious that RGA elements are large and the plant is sensitive to parameter variation. Also, according to the above properties we can guess that the corresponding condition number is large and it indicates control difficulty. We have

$$\text{cond}(G(0)) = \frac{\bar{\sigma}(G(0))}{\underline{\sigma}(G(0))} = 40.64$$

■

A result regarding the integral controllability in multivariable plants in the face of plant uncertainty is presented in (Grosdidier et al. 1985). Let the actual plant transfer function matrix and its variant be $G(s)$ and $G_P(s)$, respectively.

Theorem 7.2

Assume that the nominal plant $G(s)$ is integral controllable with the controller $C(s)$. Then, the perturbed model $G_P(s)$ with the same controller remains integral controllable if

$$\frac{\|G_P - G\|}{\|G\|} < \frac{1}{\|G\|\|G^{-1}\|} = \frac{1}{\text{cond}(G)} \tag{7.18}$$

■

For a proof of the theorem, see (Grosdidier et al. 1985). Note that equation (7.18) gives an upper bound for the maximum allowable uncertainty to retain integral controllability. In fact, the maximum relative steady state gain error should be less than the inverse of the condition number of G. Then, an integral controller exists that stabilizes both $G(s)$ and $G_P(s)$. However, this result is of limited application importance, since it is input and output scaling dependent. To apply inequality (7.18), it is therefore recommended to use the optimal condition number.

7.3 Statistical Based Approach to Uncertainty and Control Configuration Selection

The first serious work in the input-output pairing of uncertain multivariable plants is given in (Chen and Seborg 2002). In this approach, uncertainty is modeled statistically, which makes its applicability rather limited. The effects of statistical uncertainty are studied on the RGA. This analysis is useful when, plant identification methods are employed to develop the plant model from input-output data. In (Chen and Seborg 2002), equation (7.19) given below derived in (Grosdidier et al. 1985) was used to study the effect of uncertainty on the RGs

$$\frac{d\lambda_{ij}}{dg_{ij}} = \frac{(1-\lambda_{ij})\lambda_{ij}}{g_{ij}} \tag{7.19}$$

It is shown that for 2×2 plants the variance of λ_{ij} is

$$\sigma_{\lambda_{ij}}^2 \approx \sum_{k=1}^{2}\sum_{l=1}^{2}\sum_{m=1}^{2}\sum_{n=1}^{2}\left(\frac{\partial\lambda_{ij}}{\partial g_{kl}}\right)_{\hat{G}}\left(\frac{\partial\lambda_{ij}}{\partial g_{mn}}\right)_{\hat{G}}\operatorname{cov}(g_{kl},g_{mn}) \tag{7.20}$$

where $\hat{G}$ is the certain steady state plant gain matrix. It is argued in (Chen and Seborg 2002) that statistically based model uncertainty descriptions are often generated when plant identification techniques are employed to model the plant from input-output data. Using equation (7.20) and the uncertain model of the plant, it is shown that

$$(\lambda_{ij}-\hat{\lambda}_{ij})^2 \le x_{\alpha,q}^2\sigma_{\lambda_{ij}}^2 \tag{7.21}$$

where x_q^2 is the chi-square distribution with q degrees of freedom and $x_{\alpha,q}^2$ is the upper α percentage point of x_q^2, also $\hat{\lambda}_{ij}$ is the certain relative gain corresponding to the certain part of the plant. In the identification process, q is the number of parameters. Equation (7.21) can be used to compute the worst-case bounds for the uncertain steady state RGs and these confidence intervals can be extended to the dynamic RGA. The results for the 2×2 plants are extended for the general $n\times n$ multivariable plants. It is shown that

$$\frac{\partial\lambda_{ij}}{\partial g_{kl}} = \begin{cases} \dfrac{\lambda_{ij}(1-\lambda_{ij})}{g_{ij}} & i=k,\ \text{and}\ j=l \\[2ex] -\dfrac{\lambda_{ij}\lambda_{kl}}{g_{kl}} & i=k,\ \text{or}\ j=l \\[2ex] \dfrac{(-1)^{i+j+k+l}g_{ij}\det(G^{ij,kl})}{\det(G)} - \dfrac{\lambda_{ij}\lambda_{kl}}{g_{kl}} & i\ne k,\ \text{and}\ j\ne l \end{cases} \tag{7.22}$$

where $G^{ij,kl}$ denotes the sub matrix of G with rows i and k, columns j and l removed. Then, the variation bounds of uncertain relative gain for $\hat{G}_{ij} + \Delta G_{ij}$ is

$$\left|\lambda_{ij} - \hat{\lambda}_{ij}\right| \le \left|\sum_{k=1}^{n}\sum_{l=1}^{n}\left(\frac{\partial \lambda_{ij}}{\partial g_{kl}}\right)(\Delta g_{kl})\right| \tag{7.23}$$

Hence, worst-case bounds for the elements of the uncertain RGA are derived. Moreover, using these bounds, we can analyze the input-output pairing problem for uncertain multivariable plants. However, the bounds can be loose in general and even when the lower bound on the relative gain is negative, its real value may be positive for the uncertainty range.

Example 7.3.1

Consider the distillation column introduced in example 7.2.7, where its nominal dc gain is

$$\hat{G} = \begin{bmatrix} -21.6 & 1.26 \\ -2.75 & -4.28 \end{bmatrix}$$

and the corresponding RGA is

$$\hat{\Lambda} = \begin{bmatrix} 0.964 & 0.036 \\ 0.036 & 0.964 \end{bmatrix}$$

But the dc gain in the presence of statistical uncertainty with $\sigma^2_{g_{ij}} = 0.1$ is as shown in figure 7.10 and the corresponding RGA is shown in figure 7.11. Employing equation (7.20), we have $\sigma^2_{\lambda_{11}} = 0.92\times10^{-4}$ and the variance estimation certifies it as $\sigma^2_{\hat{\lambda}_{11}} = 0.97\times10^{-4}$. Now, using equation (7.21) and for $x^2_{0.05,64} = 9.14$ the confidence bounds for λ_{11} and λ_{12} shown in figure 7.12 are as follows. This shows that there is no change in the input-output pairing.

$$0.87 \le \lambda_{11} \le 1.05$$
$$0.05 \le \lambda_{12} \le 0.13$$

■

7.4 Pairing in Norm-Bounded Uncertain Plants

This section presents an input-output pairing analysis for uncertain multivariable plants. The uncertain plant is modeled with additive norm-bounded uncertainties.

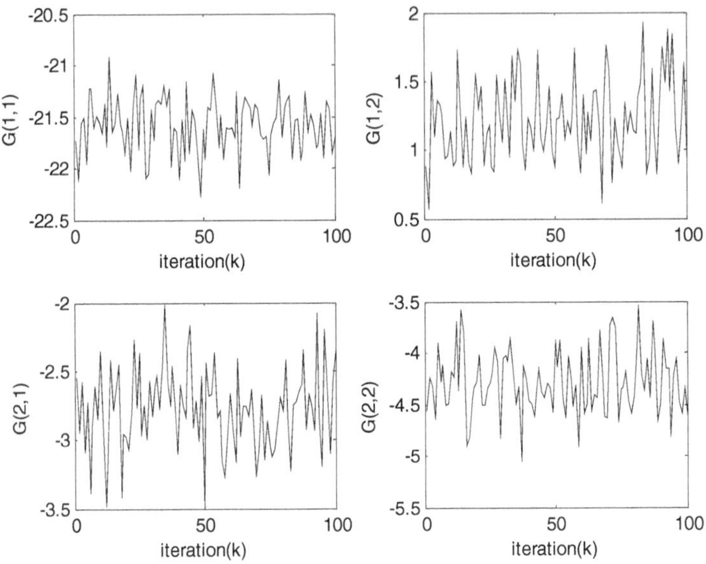

Fig. 7.10 The dc gain parameters.

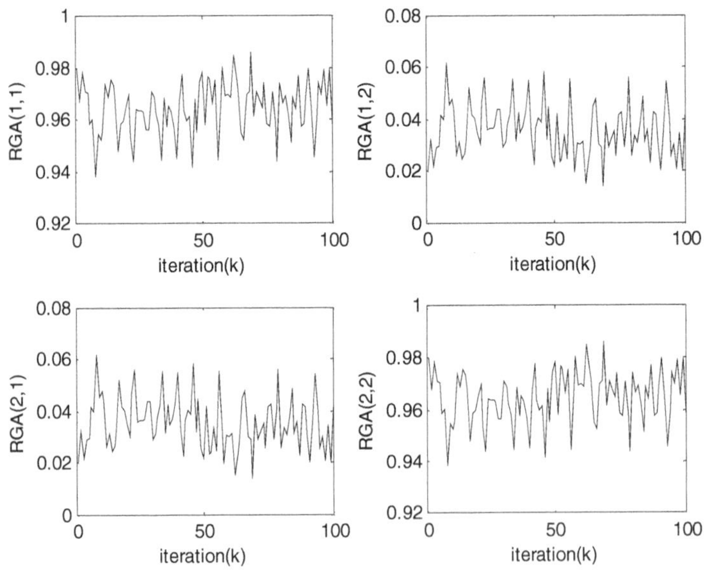

Fig. 7.11 The RGA parameters.

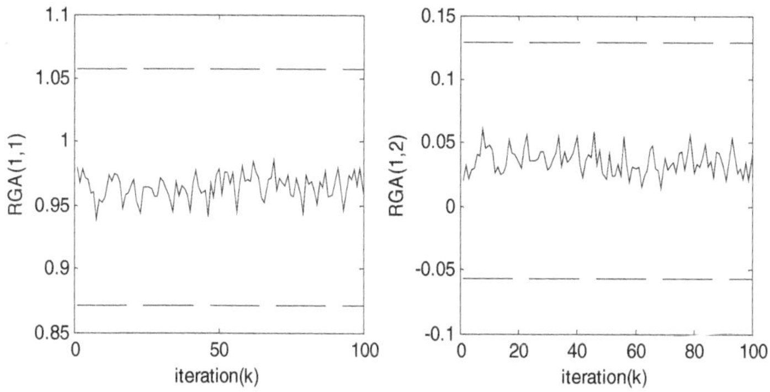

Fig. 7.12 Confidence bounds for the λ_{11} and λ_{12} variations.

The results are based on (Kariwala et al. 2006) and the key theoretical achievements are:

- Providing a signal-based representation of the Relative Gain for uncertain plants.
- Deriving a bound on the magnitude of the worst-case relative gain, calculated at steady state and also at higher frequencies.
- Deriving the necessary and sufficient conditions for the relative gain sign change in norm-bounded uncertain plants.

Let the set of steady state gain matrices of the uncertain plants Π_A be represented as

$$G_p = G + W \Delta V, \qquad \bar{\sigma}(\Delta) \le 1 \tag{7.24}$$

where G is the steady state value of the transfer function matrix of the stable, rational, linear time invariant and square nonsingular multivariable plant $G(s)$. Also, W and V are weighting matrices. Similarly, the uncertainty set is defined for other frequencies. It is assumed that the uncertain plant is functionally controllable at all frequencies and therefore its inverse exists at all frequencies. Then, the corresponding RGA for the nominal and uncertain plant are

$$\Lambda = [\lambda_{ij}(G)] = G.*G^{-T} \tag{7.25}$$

and

$$\Lambda_u = [\lambda_{ij}(G_p)] = (G + W\Delta V).*(G + W\Delta V)^{-T} \tag{7.26}$$

respectively.

It is most useful to derive an upper linear fractional transform (upper-LFT) representation of this norm-bounded uncertain plant. This will facilitate the calculation of the magnitude of worst-case relative gain.

By defining e_i as the unit column vector with its ith element equal to 1 and the remaining elements equal to 0, we can write for the whole uncertainty set, using equation (7.26) the diagonal RGA elements as

$$\lambda_{ii}(G_p) = \left(e_i^T G_p e_i\right)\left(e_i^T G_p^{-1} e_i\right) \tag{7.27}$$

Note that equation (7.24) gives

$$G_p^{-1} = (G + W\Delta V)^{-1} = \left[G\left(I + G^{-1}W\Delta V\right)\right]^{-1} = \left(I + G^{-1}W\Delta V\right)^{-1} G^{-1} \tag{7.28}$$

Now let $y = \lambda_{ii}(G_p)u$, then using equations (7.27) and (7.28), $\lambda_{ii}(G_p)$ can be represented by figure 7.13.

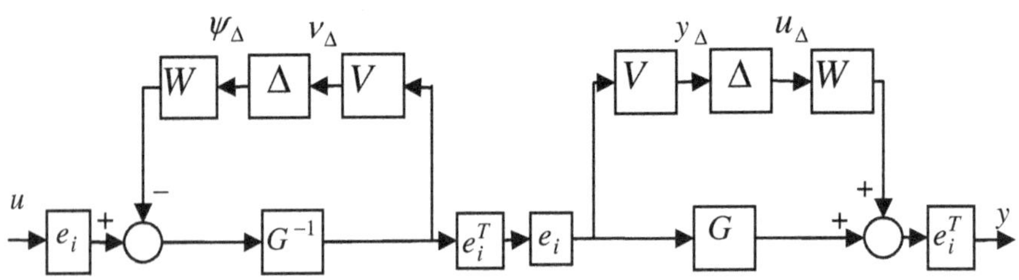

Fig. 7.13 Signal-based representation of the uncertain relative gain.

Using figure 7.13, the uncertain relative gain can be represented as an upper-LFT similar to figure 7.14, where

$$[y_\Delta^T \quad v_\Delta^T \quad y]^T = \mathbf{M}[u_\Delta^T \quad \psi_\Delta^T \quad u]^T \tag{7.29}$$

and

$$[u_\Delta^T \quad \psi_\Delta^T] = \overline{\mathbf{\Delta}}[y_\Delta^T \quad v_\Delta^T] \tag{7.30}$$

where by denoting $E = e_i e_i^T$, we have

$$\mathbf{M} = \begin{bmatrix} M_{11} & M_{12} \\ M_{21} & M_{22} \end{bmatrix} = \left[\begin{array}{cc|c} 0 & -VEG^{-1}W & VEG^{-1}e_i \\ 0 & -VG^{-1}W & VG^{-1}e_i \\ \hline e_i^T W & -e_i^T GEG^{-1}W & e_i^T GEG^{-1}e_i \end{array}\right] \tag{7.31}$$

and

$$\bar{\mathbf{\Delta}} = \begin{bmatrix} \mathbf{\Delta} & 0 \\ 0 & \mathbf{\Delta} \end{bmatrix} \tag{7.32}$$

So, the uncertain relative gain can also be represented as an upper-LFT as shown in figure 7.14. Alternatively, we have from figure 7.14

$$\lambda_{ii}(G_p) = M_{22} + M_{21}\bar{\Delta}\left(I - M_{11}\bar{\Delta}\right)^{-1} M_{12} \triangleq F_u(\mathbf{M}, \bar{\mathbf{\Delta}}) \tag{7.33}$$

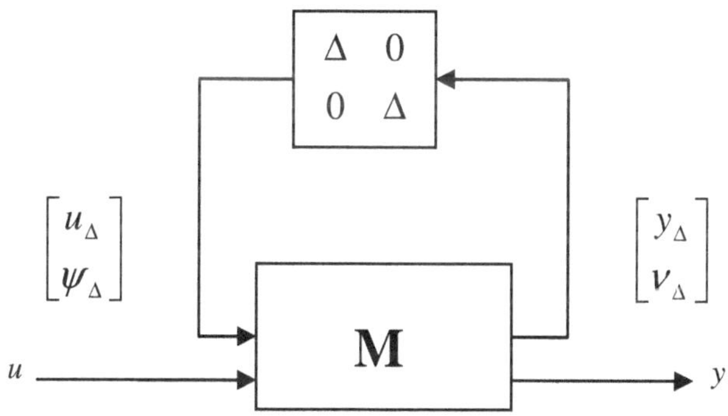

Fig. 7.14 Upper-LFT representation of uncertain relative gain.

Note 1. It is proved in (Kariwala et al. 2006) that for nonsingular G_p, the matrix $\left(I - M_{11}\bar{\Delta}\right)$ will be nonsingular for the whole allowable uncertainty set.

Note 2. For $\Delta = 0$, i.e. the nominal plant, we have $F_u(\mathbf{M}, \bar{\mathbf{\Delta}}) = M_{22} = \lambda_{ii}(G)$.

Definition 7.1. The structured singular value (SSV) of a given matrix A is defined as

$$\mu_\Delta(A) = \left(\min \bar{\sigma}(\Delta) : \det(I - A\Delta) = 0\right)^{-1} \tag{7.34}$$

and $\mu_\Delta(A) = 0$ if no allowable Δ makes $I - A\Delta$ singular.

Now assume that the following inequality holds for $\gamma \in \Re^+$

$$\mu_{\tilde{\Delta}} \begin{bmatrix} M_{11} & M_{12} \\ \gamma M_{21} & \gamma M_{22} \end{bmatrix} \le 1 \tag{7.35}$$

where

$$\tilde{\Delta} = \begin{bmatrix} \bar{\Delta} & 0 \\ 0 & \delta \end{bmatrix} \qquad |\delta| \le 1 \tag{7.36}$$

We have from the SSV definition

$$\det\left(I - \begin{bmatrix} M_{11} & M_{12} \\ \gamma M_{21} & \gamma M_{22} \end{bmatrix} \begin{bmatrix} \bar{\Delta} & 0 \\ 0 & \delta \end{bmatrix} \right) = \det\left(\begin{bmatrix} I - M_{11}\bar{\Delta} & -M_{12}\delta \\ -\gamma M_{21}\bar{\Delta} & 1 - \gamma M_{22}\delta \end{bmatrix} \right) \ne 0 \tag{7.37}$$

Expanding the last determinant gives

$$\det\left(I - M_{11}\bar{\Delta}\right) \det\left(\left(1 - \gamma M_{22}\delta\right) - \left(-\gamma M_{21}\bar{\Delta}\right)\left(I - M_{11}\bar{\Delta}\right)^{-1}\left(-M_{12}\delta\right) \right) \ne 0 \tag{7.38}$$

from note 1, we conclude that

$$\left(1 - \gamma M_{22}\delta\right) - \left(-\gamma M_{21}\bar{\Delta}\right)\left(I - M_{11}\bar{\Delta}\right)^{-1}\left(-M_{12}\delta\right) \ne 0 \tag{7.39}$$

which can be rewritten as

$$1 - \gamma\left(M_{22} + M_{21}\bar{\Delta}\left(I - M_{11}\bar{\Delta}\right)^{-1} M_{12} \right)\delta = 1 - \gamma F_u(\mathbf{M}, \bar{\Delta})\delta \ne 0 \tag{7.40}$$

Hence, equation (7.40) gives

$$1 - \gamma\lambda_{ii}(G_p)\delta \ne 0 \tag{7.41}$$

which can be written as

$$\max_{G_p \in \Pi_A} \left| \lambda_{ii}(G_p) \right| \le \gamma^{-1} \tag{7.42}$$

These results are summarized in the following proposition (Kariwala et al. 2006).

Theorem 7.3

Let G_p given by equation (7.24), be nonsingular for all the allowable uncertainty such that $\lambda_{ii}(G_p)$ in equation (7.27) is well-defined. Assume that the positive real scalar γ satisfies

$$\mu_{\tilde{\Delta}} \begin{bmatrix} M_{11} & M_{12} \\ \gamma M_{21} & \gamma M_{22} \end{bmatrix} \le 1 \tag{7.43}$$

where

$$\tilde{\Delta} = \begin{bmatrix} \Delta & 0 \\ 0 & \delta \end{bmatrix} \tag{7.44}$$

$|\delta| \le 1$ and M_{ij} $(i, j = 1, 2)$ given in (7.31). Then

$$\max_{G_p \in \Pi_A} \left| \lambda_{ii}(G_p) \right| \le \gamma^{-1} \tag{7.45}$$

■

Note 3. The magnitude of the worst-case relative gain can be found by finding the largest γ satisfying the above theorem.

Note 4. In practice, computation of γ is done by equating both sides of the inequality (7.43) and using the bisection algorithm.

Example 7.4.1

Consider the steady state transfer function matrix of the Wood and Berry distillation column introduced in example 2.2.1

$$G = G(0) = \begin{bmatrix} 12.8 & -18.9 \\ 6.6 & -19.4 \end{bmatrix} \tag{7.46}$$

and the corresponding uncertainty is defined as

$$W = \begin{bmatrix} |G(1,1)| & |G(1,2)| & 0 & 0 \\ 0 & 0 & |G(2,1)| & |G(2,2)| \end{bmatrix}$$

$$V = \begin{bmatrix} \alpha & 0 & \alpha & 0 \\ 0 & \alpha & 0 & \alpha \end{bmatrix}^T \tag{7.47}$$

$$\Delta = diag(\delta_{ij}) \qquad |\delta_{ij}| \le 1$$

According to the above theorem and note 4, for $\alpha = 0.01$ the inequality (7.43) results in $\gamma = 0.4677$, hence $\max_{G_p} |\lambda_{11}(G_p)| \le 2.1382$. Also, for $\alpha = 0.1$ the magnitude of the worst-case for element (1,1) of the perturbed RGA is $\max_{G_p} |\lambda_{11}(G_p)| \le 4.8972$.

■

Note 5. Integrity in the face of plant uncertainty or *robust integrity* is an important issue which has recently been addressed. The plant $G(s)$ has robust integrity if there exists a diagonal controller $\hat{K}(s) = HK(s)$ with integral action, which stabilizes $G_p(s)$ for all the allowable uncertainties and $H \in \mathrm{H}$, where $\mathrm{H} = \{H = diag(h_i) | h_i = \{0,1\}, i = 1, \cdots, m\}$. For some preliminary results and discussions see (Kariwala et al. 2006).

Note 6. The critical issue of sign changes of the RG of uncertain multivariable plants is also addressed (Kariwala et al. 2006). Necessary and sufficient conditions for sign change of RGs, using the structured singular value (SSV) framework are derived. It is shown that for G_p satisfying the conditions of the theorem 7.3, $\lambda_{ii}(G_p)$ changes sign over the uncertainty set Π_A if and only if

$$\mu_{\bar{\Delta}}(\mathbf{N}) > 1 \tag{7.48}$$

where

$$\mathbf{N} = \begin{bmatrix} -\lambda_{ii}^{-1}(G)VEG^{-1}EW & -VEG^{-1}\left(I - \lambda_{ii}^{-1}(G)EGEG^{-1}\right)W \\ -\lambda_{ii}^{-1}(G)VG^{-1}EW & -VG^{-1}\left(I - \lambda_{ii}^{-1}(G)EGEG^{-1}\right)W \end{bmatrix} \tag{7.49}$$

Note 7. There are two main drawbacks associated with the method. First, its implementation in large scale or complex systems is computationally costly. Second, it cannot detect the appropriate input-output pair when there is no sign change in the relative gains and most of them are positive.

In (Moaveni and Khaki-Sedigh 2007b), another presentation of the uncertain RG is provided. Where, the uncertain RG $\lambda_{ii}(G_p)$ is represented as the certain RG $\lambda_{ii}(G)$ and an additive uncertainty δ_{ii}, as shown in figure 7.15.

Note that each uncertain RG shown in figure 7.13 can be represented as

$$\begin{aligned}\lambda_{ii}(G_p) &= e_i^T(G+\Delta_T)e_i e_i^T(G+\Delta_T)^{-1}e_i \\ &= e_i^T(G+\Delta_T)E_i(G+\Delta_T)^{-1}e_i\end{aligned} \tag{7.50}$$

where Δ_T is unstructured uncertainty and the corresponding certain relative gain can be written as

$$\lambda_{ii}(G) = e_i^T G e_i e_i^T G^{-1} e_i = e_i^T G E_i G^{-1} e_i \tag{7.51}$$

hence

$$\delta_{ii} = \lambda_{ii}(G_p) - \lambda_{ii}(G) = e_i^T(G+\Delta_T)E_i(G+\Delta_T)^{-1}e_i - e_i^T G E_i G^{-1} e_i \tag{7.52}$$

and

$$\delta_{ii} = e_i^T G\left\{G^{-1}\Delta_T E_i - E_i G^{-1}\Delta_T\right\}(G+\Delta_T)^{-1}e_i \tag{7.53}$$

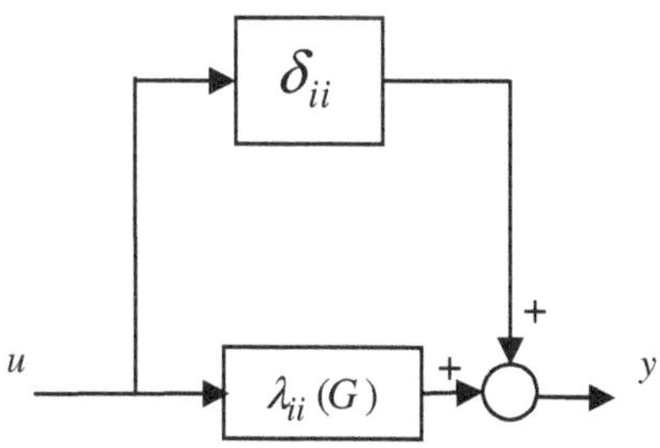

Fig. 7.15 Representation of the uncertain relative gain.

Now, by using the 2-norm and standard norm properties we have

$$\begin{aligned}\|\delta_{ii}\| &= \left\|e_i^T G\left\{G^{-1}\Delta_T E_i - E_i G^{-1}\Delta_T\right\}(G+\Delta_T)^{-1}e_i\right\| = \\ &= \bar{\sigma}(e_i^T G\left\{G^{-1}\Delta_T E_i - E_i G^{-1}\Delta_T\right\}(G+\Delta_T)^{-1}e_i)\end{aligned} \tag{7.54}$$

and

$$\|\delta_{ii}\| \le \frac{\bar{\sigma}(e_i^T G)\bar{\sigma}(G^{-1}\Delta_T E_i - E_i G^{-1}\Delta_T)}{\underline{\sigma}(G+\Delta_T)} \tag{7.55}$$

where if $\bar{\sigma}(\Delta_T) < \underline{\sigma}(G)$ then inequality (7.55) can be rewritten as

$$\begin{aligned}\|\delta_{ii}\| &\le \frac{\bar{\sigma}(e_i^T G)\left\{\bar{\sigma}(G^{-1}\Delta_T E_i) + \bar{\sigma}(E_i G^{-1}\Delta_T)\right\}}{\underline{\sigma}(G) - \bar{\sigma}(\Delta_T)} \\ &\le \frac{2\bar{\sigma}(e_i^T G)\bar{\sigma}(E_i G^{-1})\bar{\sigma}(\Delta_T)}{\underline{\sigma}(G) - \bar{\sigma}(\Delta_T)} = \delta_{ii}^{up}\end{aligned} \tag{7.56}$$

So, the variation bound of uncertain relative gain is

$$\lambda_{ii} - \delta_{ii}^{up} \le \lambda_{ii}^{u} \le \lambda_{ii} + \delta_{ii}^{up} \tag{7.57}$$

The above results are summarized in the following theorem (Moaveni and Khaki-Sedigh 2007b).

Theorem 7.4

Consider the linear multivariable plant G and the corresponding RGA $\Lambda = [\lambda_{ij}],\ i, j = 1, 2, \ldots, m$ where in the presence of additive uncertainty the uncertain RGA is

$$\Lambda(G_p) = [\lambda_{ij}(G_p)] = [\lambda_{ij} + \delta_{ij}] = (G+\Delta_T).*(G+\Delta_T)^{-T} \tag{7.58}$$

Therefore, if $\bar{\sigma}(\Delta_T) < \underline{\sigma}(G)$ then

$$\|\delta_{ij}\| \le \frac{2\bar{\sigma}(e_i^T G)\bar{\sigma}(E_j G^{-1})\bar{\sigma}(\Delta_T)}{\underline{\sigma}(G) - \bar{\sigma}(\Delta_T)} = \delta_{ij}^{up} \tag{7.59}$$

and the variation bound of uncertain relative gains is

$$\lambda_{ij} - \delta_{ij}^{up} \leq \lambda_{ij}^{u} \leq \lambda_{ij} + \delta_{ij}^{up} \tag{7.60}$$

■

Note 1. This result decreases the computational load in large-scale uncertain plants. But, the computed bounds will be loose. If there is no overlap between the variation bounds of the same row and the same column in the uncertain RGA, the nominal input-output pairing remains valid for all parameter variations. Otherwise, the nominal input-output pairing may change due to parameter variations.

Example 7.4.2

Consider the Wood and Berry distillation column introduced in example 7.4.1 with a similar uncertainty. Using equations (7.24), (7.47) and (7.50) we have

$$\Delta_T = W\,\Delta V = \alpha \begin{bmatrix} \delta_{11}\left|G(1,1)\right| & \delta_{12}\left|G(1,2)\right| \\ \delta_{21}\left|G(2,1)\right| & \delta_{22}\left|G(2,2)\right| \end{bmatrix}$$

and for $\left|\delta_{ij}\right| \leq 1$, $\alpha = 0.01$ and $\alpha = 0.1$ condition of theorem 7.4 is always satisfied

$\bar{\sigma}(\Delta_T) < \underline{\sigma}(G)$

Hence, for $\alpha = 0.01$ the upper bound for the variation bound of element (1,1) of the RGA is

$\left|\lambda_{11}(G_p) - \lambda_{11}\right| \leq \delta_{ij}^{up} = 0.81$

Therefore

$1.2 \leq \lambda_{11}(G_p) \leq 2.82$

where this inequality indicates that the above uncertainty can not change the appropriate input-output pair. But, for $\alpha = 0.1$ we have

$\delta_{11}^{up} = 29.71$

and it is obvious that there is a possible input-output pairing change.

■

Comparing the results of examples 7.4.1 and 7.4.2 show that in the first approach, only the worst-case magnitudes of RGs are derived, and if no sign change occurs for the RGs, we cannot deduce any new information regarding the input-output pairing analysis. But, the second approach introduces the variation bounds of the RGs due to the unstructured uncertainty, where it can be used to deduce new information regarding the input-output pairing analysis. However, the bounds are loose in comparison with the first method.

7.5 Pairing of Uncertain Multivariable Plants Using the DIOPM

In this section, after presenting a very short summary of some norm definitions and properties used later, uncertainty in state space models is considered. The results are derived from (Moaveni and Khaki-Sedigh 2008).

7.5.1 Some Norm Properties

Consider the matrices $M = [m_{ij}]$ and $N = [n_{ij}] \in \Re^{n \times n}$. Then the following norm definitions are given.

Frobenius norm. Frobenius norm (or Euclidean norm) denoted by $\|M\|_F$ is the square root of the sum of the squared elements.

$$\|M\|_F = \sqrt{\sum_{i,j} m_{ij}^2} = \sqrt{\mathrm{trace}(M^T M)}$$

2-norm. 2-norm denoted by $\|M\|$ is the largest singular value of the corresponding matrix.

$$\|M\| = \bar{\sigma}(M) = \sqrt{\lambda_{\max}(M^T M)}$$

where $\bar{\sigma}$ denotes the largest singular value.

Norm relationships. Following important norm relationships are used later.

$$\begin{aligned}
&\|MN\| \le \|M\| \|N\| \\
&\big|\|M\| - \|N\|\big| \le \|M + N\| \le \|M\| + \|N\| \\
&\|M\| \le \|M\|_F \le \sqrt{n}\|M\| \\
&\|vec(M)\| = \|M\|_F \\
&\|MN\|_F \le \|M\|_F \|N\|_F
\end{aligned} \tag{7.61}$$

where "vec" operator stacks the columns of a matrix into a long vector.

7.5.2 The Main Result

In this section the DIOPM is used for input-output pairing analysis. The DIOPM can be computed using the cross-Gramian matrix given by equation (5.70). The following uncertain multivariable plant is considered

$$\begin{aligned} \dot{X} &= (A+\Delta A)X + (B+\Delta B)U = (A+\Delta A)X + \sum_{i=1}^{m}(b_{*i}+\Delta b_{*i})u_i \\ Y &= (C+\Delta C)X \Rightarrow Y = \left[y_j\right] = \left[(c_{j*}+\Delta c_{j*})X\right] \end{aligned} \tag{7.62}$$

where b_{*i} and c_{j*} are the ith column of B and the jth row of C respectively. Also, ΔA is the additive uncertainty in matrix A, Δb_{*i} is the additive uncertainty in the ith column of B and Δc_{j*} is the additive uncertainty in the jth row of C. Also, it is assumed that the upper bounds of $\|\Delta A\|$, $\|\Delta b_{*i}\|$ and $\|\Delta c_{j*}\|$ are known.

To compute the DIOPM elements $\varphi_{ij} = \bar{\sigma}(W_{co}^{ij})$, the matrix W_{co}^{ij} should be computed from the following equations

$$W_{co}^{ij}A + AW_{co}^{ij} = -b_{*i}c_{j*} \tag{7.63}$$

In the presence of additive uncertainty, the uncertain DIOPM is as

$$\varphi_{ij}^{u} = \sigma(W_{co}^{ij} + \Delta^{ij}) = \max(|\lambda(W_{co}^{ij} + \Delta^{ij})|) \tag{7.64}$$

where

$$(W_{co}^{ij} + \Delta^{ij})(A+\Delta A) + (A+\Delta A)(W_{co}^{ij} + \Delta^{ij}) = -(b_{*i}+\Delta b_{*i})(c_{j*}+\Delta c_{j*}) \tag{7.65}$$

and using (7.61) it is obvious that

$$\|W_{co}^{ij}\| - \|\Delta^{ij}\| \le \|W_{co}^{ij} + \Delta^{ij}\| \le \|W_{co}^{ij}\| + \|\Delta^{ij}\| \tag{7.66}$$

where it shows the variation bounds of the elements of uncertain DIOPM and it can be summarized as

$$\left|\left(\|W_{co}^{ij} + \Delta^{ij}\| - \|W_{co}^{ij}\|\right)\right| \le \|\Delta^{ij}\| \tag{7.67}$$

This inequality introduces an upper bound for the element variation in DIOPM due to unstructured uncertainty in the plant. So, the pairing analysis of the uncertain plant can be done by computing the upper bound of equation (7.67). To compute this upper bound, equation (7.65) is rewritten as

$$\begin{aligned} W_{co}^{ij} \cdot A + W_{co}^{ij} \cdot \Delta A + \Delta^{ij} \cdot A + \Delta^{ij} \cdot \Delta A + A \cdot W_{co}^{ij} + A \cdot \Delta^{ij} + \Delta A \cdot W_{co}^{ij} + \\ + \Delta A \cdot \Delta^{ij} = -b_{*i} c_{j*} - b_{*i} \Delta c_{j*} - \Delta b_{*i} c_{j*} - \Delta b_{*i} \Delta c_{j*} \end{aligned} \tag{7.68}$$

Using equations (7.63) and (7.68), it can be shown that

$$\begin{aligned} W_{co}^{ij} \cdot \Delta A + \Delta^{ij} \cdot A + \Delta^{ij} \cdot \Delta A + A \cdot \Delta^{ij} + \Delta A \cdot W_{co}^{ij} + \Delta A \cdot \Delta^{ij} = \\ -b_{*i} \Delta c_{j*} - \Delta b_{*i} c_{j*} - \Delta b_{*i} \Delta c_{j*} \end{aligned} \tag{7.69}$$

and

$$\begin{aligned} A \cdot \Delta^{ij} + \Delta A \cdot \Delta^{ij} + \Delta^{ij} \cdot A + \Delta^{ij} \cdot \Delta A = \\ -b_{*i} \Delta c_{j*} - \Delta b_{*i} c_{j*} - \Delta b_{*i} \Delta c_{j*} - \Delta A \cdot W_{co}^{ij} - W_{co}^{ij} \cdot \Delta A \end{aligned} \tag{7.70}$$

Dropping the second order uncertainty terms, equation (7.70) gives

$$\begin{aligned} (I_n \otimes A + A^T \otimes I_n) vec(\Delta^{ij}) = \\ -vec(b_{*i} \Delta c_{j*} + \Delta b_{*i} c_{j*} + \Delta b_{*i} \Delta c_{j*} + \Delta A \cdot W_{co}^{ij} + W_{co}^{ij} \cdot \Delta A) \end{aligned} \tag{7.71}$$

where "$\otimes$" denotes the Kronecker product. Now we can compute the $vec(\Delta^{ij})$ as follows

$$\begin{aligned} vec(\Delta^{ij}) = -(I_n \otimes A + A^T \otimes I_n)^{-1} \times \\ vec(b_{*i} \Delta c_{j*} + \Delta b_{*i} c_{j*} + \Delta b_{*i} \Delta c_{j*} + \Delta A \cdot W_{co}^{ij} + W_{co}^{ij} \cdot \Delta A) \end{aligned} \tag{7.72}$$

By taking the 2-norm

$$\begin{aligned} \left\| vec(\Delta^{ij}) \right\| = \Big\| (I_n \otimes A + A^T \otimes I_n)^{-1} \times \\ vec(b_{*i} \Delta c_{j*} + \Delta b_{*i} c_{j*} + \Delta b_{*i} \Delta c_{j*} + \Delta A \cdot W_{co}^{ij} + W_{co}^{ij} \cdot \Delta A) \Big\| \end{aligned} \tag{7.73}$$

and using the norm properties, we have

$$\begin{aligned} \left\| \Delta^{ij} \right\|_F = \left\| vec(\Delta^{ij}) \right\| \le \left\| (I_n \otimes A + A^T \otimes I_n)^{-1} \right\| \times \\ \left\| vec(b_{*i} \Delta c_{j*} + \Delta b_{*i} c_{j*} + \Delta b_{*i} \Delta c_{j*} + \Delta A \cdot W_{co}^{ij} + W_{co}^{ij} \cdot \Delta A) \right\| \end{aligned} \tag{7.74}$$

hence

$$\begin{aligned}\left\|\Delta^{ij}\right\| \leq \left\|\Delta^{ij}\right\|_F \leq \left\|(I_n \otimes A + A^T \otimes I_n)^{-1}\right\| \times \\ \times \left\| vec(b_{*i}\Delta c_{j*}) + vec(\Delta b_{*i} c_{j*}) + vec(\Delta b_{*i}\Delta c_{j*}) + vec(\Delta A \cdot W_{co}^{ij}) + vec(W_{co}^{ij} \cdot \Delta A)\right\|\end{aligned} \tag{7.75}$$

Using the norm inequalities

$$\begin{aligned}\left\|\Delta^{ij}\right\| \leq \left\|(I_n \otimes A + A^T \otimes I_n)^{-1}\right\| \times \\ \times \left(\left\|vec(b_{*i}\Delta c_{j*})\right\| + \left\|vec(\Delta b_{*i} c_{j*})\right\| + \left\|vec(\Delta b_{*i}\Delta c_{j*})\right\| + \left\|vec(\Delta A \cdot W_{co}^{ij})\right\| + \left\|vec(W_{co}^{ij} \cdot \Delta A)\right\| \right)\end{aligned} \tag{7.76}$$

and

$$\begin{aligned}\left\|\Delta^{ij}\right\| \leq \left\|(I_n \otimes A + A^T \otimes I_n)^{-1}\right\| \times \\ \left(\left\|b_{*i}\Delta c_{j*}\right\|_F + \left\|\Delta b_{*i} c_{j*}\right\|_F + \left\|\Delta b_{*i}\Delta c_{j*}\right\|_F + \left\|\Delta A \cdot W_{co}^{ij}\right\|_F + \left\|W_{co}^{ij} \cdot \Delta A\right\|_F \right)\end{aligned} \tag{7.77}$$

Also, it is obvious that

$$\begin{aligned}\left\|\Delta^{ij}\right\| \leq \left\|(I_n \otimes A + A^T \otimes I_n)^{-1}\right\| \times \\ \times (\left\|b_{*i}\right\|_F \left\|\Delta c_{j*}\right\|_F + \left\|\Delta b_{*i}\right\|_F \left\|c_{j*}\right\|_F + \left\|\Delta b_{*i}\right\|_F \left\|\Delta c_{j*}\right\|_F + \left\|\Delta A\right\|_F \left\|W_{co}^{ij}\right\|_F + \left\|W_{co}^{ij}\right\|_F \left\|\Delta A\right\|_F)\end{aligned} \tag{7.78}$$

Finally, we have

$$\begin{aligned}\left\|\Delta^{ij}\right\| \leq \left\|(I_n \otimes A + A^T \otimes I_n)^{-1}\right\| \times \\ \left(2n\left\|\Delta A\right\| \left\|W_{co}^{ij}\right\| + \left\|b_{*i}\right\| \left\|\Delta c_{j*}\right\| + \left\|\Delta b_{*i}\right\| \left\|c_{j*}\right\| + \left\|\Delta b_{*i}\right\| \left\|\Delta c_{j*}\right\| \right)\end{aligned} \tag{7.79}$$

So, in (7.79) the upper bound of the element variations in uncertain DIOPM is computed. The above results can be summarized in the following theorem (Moaveni and Khaki-Sedigh 2008).

Theorem 7.5

For the uncertain linear multivariable plant given by equation (7.62), if $\Phi^u = [\max\{|\lambda(W_{co}^{ij} + \Delta^{ij})|\}]$ is the corresponding uncertain DIOPM, and Δ^{ij} is the variation of matrix W_{co}^{ij} in the presence of uncertainty, then the upper bound of $\left\|\Delta^{ij}\right\|$ is

$$\begin{aligned}\left\|\Delta^{ij}\right\| \leq \left\|(I_n \otimes A + A^T \otimes I_n)^{-1}\right\| \times \\ \left(2n\left\|\Delta A\right\| \left\|W_{co}^{ij}\right\| + \left\|b_{*i}\right\| \left\|\Delta c_{j*}\right\| + \left\|\Delta b_{*i}\right\| \left\|c_{j*}\right\| + \left\|\Delta b_{*i}\right\| \left\|\Delta c_{j*}\right\| \right)\end{aligned} \tag{7.80}$$

and the upper bound of DIOPM elements variations in the presence of additive uncertainty is

$$\left|\left(\left\|W_{co}^{ij}+\Delta^{ij}\right\|-\left\|W_{co}^{ij}\right\|\right)\right|\leq\left\|\Delta^{ij}\right\| \tag{7.81}$$

■

The above theorem provides an upper bound for the element variations in the DIOPM, due to the unstructured uncertainty in the state space model of plant. An algorithm is now proposed that using the above theorem can show the possible input-output pairing changes resulting from the model uncertainty.

Step 1. Compute the DIOPM for the nominal plant using (5.70).

Step 2. Calculate the variation bounds of the DIOPM elements variations using (7.80).

Step 3. Consider the following two cases:

- If there is no overlap between variation bounds of the same row and the same column in the DIOPM, the nominal input-output pairing remains valid for all parameter variations.
- If there is an overlap between variation bounds of the same row or the same column in the DIOPM, the nominal input-output pairing may change due to parameter variations.

Example 7.5.1

Consider the Quadruple-tank with the following nonlinear state space model given in example 2.3.1

$$\dot{h}_1=-\frac{a_1}{A_1}\sqrt{2gh_1}+\frac{a_3}{A_1}\sqrt{2gh_3}+\frac{\gamma_1 k_1}{A_1}u_1$$

$$\dot{h}_2=-\frac{a_2}{A_2}\sqrt{2gh_2}+\frac{a_4}{A_2}\sqrt{2gh_4}+\frac{\gamma_2 k_2}{A_2}u_2$$

$$\dot{h}_3=-\frac{a_3}{A_3}\sqrt{2gh_3}+\frac{(1-\gamma_2)k_2}{A_3}u_2$$

$$\dot{h}_4=-\frac{a_4}{A_4}\sqrt{2gh_4}+\frac{(1-\gamma_1)k_1}{A_4}u_1$$

To analyze the input-output pairing using the above approach, the nonlinear plant is modeled as a linear plant with structured uncertainty. Assume that the Quadruple-tank has the following physical constants:

$A_1 = A_3 = 28 \ , \ A_2 = A_4 = 32 \ (cm^2)$

$a_1 = a_3 = 0.071 \ , \ a_2 = a_4 = 0.057 \ (cm^2)$

$k_c = 0.50 \ (V\ /cm)$

$g = 981 \ (cm\ /s^2)$

$k_1 = k_2 = 2.9$

The parameters that vary and can change the appropriate input-output pair of the plant are γ_1 and γ_2. Two cases are considered for the uncertainty analysis. The nominal parameters $\gamma_1 = 0.42$, $\gamma_2 = 0.32$, $h_1 = 13.64$, $h_2 = 16.55$, $h_3 = 1.91$ and $h_4 = 1.77$ give

$$A = \begin{bmatrix} -0.0152 & 0 & 0.0406 & 0 \\ 0 & -0.0097 & 0 & 0.0297 \\ 0 & 0 & -0.0406 & 0 \\ 0 & 0 & 0 & -0.0297 \end{bmatrix}$$

$$B = \begin{bmatrix} 0.0500 & 0 \\ 0 & 0.0335 \\ 0 & 0.0814 \\ 0.0604 & 0 \end{bmatrix}, \ C = \begin{bmatrix} 0.5 & 0 & 0 & 0 \\ 0 & 0.5 & 0 & 0 \end{bmatrix}$$

Case 1. The parameter variations are given as

$0.38 \le \gamma_1 \le 0.46$

$0.28 \le \gamma_2 \le 0.36$

It was shown in example 2.3.1 that for $\gamma_1 + \gamma_2 < 1$, $(u_1 - y_2, u_2 - y_1)$ will be an appropriate input-output pair. So, for the above variation bounds of γ_1 and γ_2, $(u_1 - y_2, u_2 - y_1)$ is the appropriate input-output pair. Using (5.70), the Φ matrix for $\gamma_1 = 0.42$ and $\gamma_2 = 0.32$ is

$$\Phi = \begin{bmatrix} 0.8212 & 1.5642 \\ 1.8051 & 0.8637 \end{bmatrix}$$

which certifies the result in example 2.3.1. In the presence of the above variations, we have

$$\|\Delta b_{*1}\| \le 0.0063, \quad \|\Delta b_{*2}\| \le 0.0064$$

and it is obvious that

$$\|\Delta c_{j*}\| = 0, \quad j = 1,2$$

so

$$\|\Delta^{11}\| \le 0.2808, \quad \|\Delta^{12}\| \le 0.2825$$
$$\|\Delta^{21}\| \le 0.2808, \quad \|\Delta^{22}\| \le 0.2825$$

Therefore, the variation bounds of the Φ^u matrix elements are

$$\Phi^u = \begin{bmatrix} 0.5404 \le \bar{\sigma}(W_{co}^{11} + \Delta^{11}) \le 1.1020 & 1.2817 \le \bar{\sigma}(W_{co}^{12} + \Delta^{12}) \le 1.8467 \\ 1.5243 \le \bar{\sigma}(W_{co}^{21} + \Delta^{21}) \le 2.0859 & 0.5811 \le \bar{\sigma}(W_{co}^{22} + \Delta^{22}) \le 1.1462 \end{bmatrix}$$

Matrix Φ^u shows that for the assumed uncertainty, $(u_1 - y_2, u_2 - y_1)$ remains the appropriate input-output pair, because the lower bound of φ_{21}^u is greater than the upper bounds of φ_{11}^u and φ_{22}^u, respectively.

Case 2. The parameter variations are given as

$$0.32 \le \gamma_1 \le 0.52$$
$$0.12 \le \gamma_2 \le 0.52$$

In example 2.3.1, it is shown that for the above γ_1 and γ_2 variation bounds, there is a chance that the appropriate pair changes form $(u_1 - y_2, u_2 - y_1)$ to $(u_1 - y_1, u_2 - y_2)$. In this case

$$\|\Delta b_{*1}\| \le 0.0158, \quad \|\Delta b_{*2}\| \le 0.0318$$

and if (7.80) is applied

$$\|\Delta_{11}\| \le 0.7021, \quad \|\Delta_{12}\| \le 1.4125$$
$$\|\Delta_{21}\| \le 0.7021, \quad \|\Delta_{22}\| \le 1.4125$$

Then

$$\Phi^u = \begin{bmatrix} 0.1192 \le \bar{\sigma}(W_{co}^{11} + \Delta^{11}) \le 1.5233 & 0.1517 \le \bar{\sigma}(W_{co}^{12} + \Delta^{12}) \le 2.9768 \\ 1.1031 \le \bar{\sigma}(W_{co}^{21} + \Delta^{21}) \le 2.5072 & 0 \le \bar{\sigma}(W_{co}^{22} + \Delta^{22}) \le 2.2762 \end{bmatrix}$$

It is obvious that the upper bounds of φ_{11}^u and φ_{22}^u are greater than the lower bounds of φ_{12}^u and φ_{21}^u. There is, then a possible input-output pairing change.

■

7.6 Adaptive Input-Output Pairing of Unknown Multivariable Plants

Modeling uncertainties and their effects on the control configuration were discussed in the previous sections. However, the proposed techniques do not give a direct answer to the following two important questions. Consider an uncertain or unknown and time varying multivariable plant. If the extent of uncertainty or the resultant plant variations are large enough, the appropriate input-output pair may endure a change during the plant operation. Following the pairing alteration, the closed loop plant under decentralized control may therefore render unstable. The two key questions are

1. When does a parameter change results in a pairing alteration?
2. What is the new pairing after the parameter change?

To answer these questions (Khaki-Sedigh and Moaveni 2003b), and (Moaveni and Khaki-Sedigh 2007b) proposed an adaptive pairing procedure and a reconfigurable decentralized control strategy. In this section, the adaptability characteristics of the RGA is shown and Recursive Least Squares (RLS) identifiers are used to provide an updated version of the RGA. The RGA pairing rules are used to find the appropriate control configuration at each iteration. The pairing alteration can then be identified and the new pairing will be proposed.

Let the linear model of the multivariable plant at an operating point be described by the $m \times m$ transfer function matrix $G(s)$. To implement the adaptive pairing strategy, the equivalent discrete-time model with a zero order hold is derived, we have $Y(z) = G(z)U(z)$. The plants under consideration can be readily represented on the discrete time set $T_T = \{0, T, 2T, \ldots, kT, \ldots\}$ where T is the sampling time, by means of an autoregressive difference equation of the form

$$y_k + A_1 y_{k-1} + \ldots + A_n y_{k-N} = B_1 u_{k-1} + \ldots + B_N u_{k-N} \tag{7.82}$$

where the matrices $A_i \in \Re^{m\times m}$ and $B_i \in \Re^{m\times m}$ ($i = 1, 2, \ldots, N$) are the parameters of the Nth order model. These parameters of the Auto Regressive Moving

Average (ARMA) model can be conveniently estimated by implementing in real time, the RLS parameter estimation algorithm.

The key observation is that the steady-state gain matrix of the multivariable plant can be calculated from equation (7.83), as

$$G_0 = G(1) = (I + A_1 + ... + A_N)^{-1}(B_1 + B_2 + ... + B_N) \tag{7.83}$$

Hence, by invoking the certainty equivalence principle, the on-line RGA of the plant is obtained by the following procedure:

Step 1. Use the RLS algorithm to obtain the estimates of A_i and B_i matrices. Let the plant ARMA model be rewritten in the regression form $y(t) = \phi^T(t-1)\theta$, where $\phi^T(t-1)$ is the regressor and θ is the parameter vector. The RLS algorithm is summarized as follows:

$$\hat{\theta}(t) = \hat{\theta}(t-1) + K(t)[y(t) - \phi^T(t)\hat{\theta}(t-1)]$$
$$K(t) = P(t-1)\phi(t)[\lambda I + \phi^T(t)P(t-1)\phi(t)]^{-1}$$
$$P(t) = [I - K(t)\phi^T(t)]P(t-1)/\lambda$$

where $\hat{\theta}$ is the estimated parameter vector, $K(t)$ is the Kalman gain, $P(t)$ is the covariance matrix, and λ is the forgetting factor. For details and derivations, see for example (Astrom and Wittenmark 1995).

Step 2. Calculate G_0 from equation (7.83) using the estimated parameters from the RLS at each iteration.

Step 3. Calculate the RGA at every sampling time using equation (2.9).

Step 4. Evaluate the present input-output pairing and detect any possible pairing changes. This can be performed using the standard RGA pairing rules presented in chapter 2, and the results of examples 7.2.3 and 7.2.5 for 2×2 and 3×3 multivariable plants.

Example 7.6.1

Consider the following transfer function matrix introduced in example 2.3.2

$$G(s) = \frac{(1-s)}{(1+5s)^2}\begin{bmatrix} 1 & -4.19 & -25.96 \\ 6.19 & 1 & -25.96 \\ 1 & 1 & 1 \end{bmatrix}$$

and its corresponding RGA as

$$\Lambda = \begin{bmatrix} 1.00 & 5.00 & -5.00 \\ -5.00 & 1.00 & 5.00 \\ 5.00 & -5.00 & 1.00 \end{bmatrix}$$

Also, the corresponding pulse transfer function with $T = 1$ second is

$$G(z) = \frac{1}{z^2 - 1.637z + 0.6703} \times \begin{bmatrix} -0.015z + 0.048 & 0.064z - 0.201 & 0.395z - 1.248 \\ -0.095z + 0.298 & -0.015z + 0.048 & 0.395z - 1.248 \\ -0.015z + 0.048 & -0.015z + 0.048 & -0.015z + 0.048 \end{bmatrix} \tag{7.84}$$

If the two parameters of $G(z)$ be perturbed as

$$G_1(z) = \frac{1}{z^2 - 1.637z + 0.6703} \times \begin{bmatrix} -0.015z + 0.048 & 0.064z - 0.201 & 0.395z - 1.248 \\ -0.095z + \boxed{1.25 \times 0.298} & -\boxed{0.75 \times 0.015}z + 0.048 & 0.395z - 1.248 \\ -0.015z + 0.048 & -0.015z + 0.048 & -0.015z + 0.048 \end{bmatrix} \tag{7.85}$$

the corresponding RGA is

$$\Lambda_1 = \begin{bmatrix} -1.4072 & -7.4944 & 9.9016 \\ 9.5660 & -1.5635 & -7.0024 \\ -7.1588 & 10.0579 & -1.8991 \end{bmatrix} \tag{7.86}$$

Note that the input-output pairing proposed by Λ is not valid in the case of Λ_1. Figure 7.16 shows the convergence of the RGA parameters while its parameters are changed in $t = 500$ (sec). This figure clearly indicates that a new input-output pairing ought to be adopted by the decentralized control methodology.

■

7.6.1 *Reconfigurable Architecture of the Decentralized Controller*

It was shown in the previous section that the unknown, uncertain or time varying plant parameters or the faults encountered in complex multivariable plants can lead to major changes in the plant structure and hence its input-output pairing.

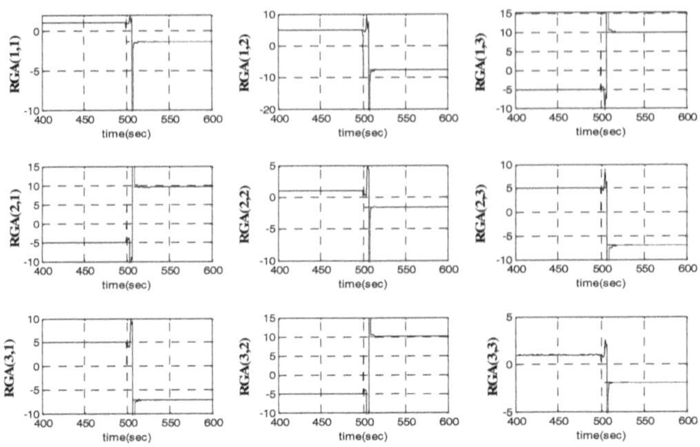

Fig. 7.16 Convergence of the RGA parameters.

These changes seriously deteriorate the closed loop decentralized performance and can easily lead to closed-loop instability. To overcome this problem, a reconfigurable decentralized controller is proposed. This employs an on-line RGA calculation along with an updated input-output selection at each sampling time. The reconfigurable control procedure is as below:

Step 1. Obtain an on-line update of the RGA matrix and detect any pairing changes as proposed in the previous sections.

Step 2. Decentralized controller is designed using an appropriate control strategy with a new input-output pair, if necessary. An actuator selection is made and the control signal is implemented through the appropriate actuator.

In order to implement the above ideas in a decentralized control structure, let the multivariable plant be described by the following transfer function matrix

$$G(z)=\frac{1}{a(z)}\begin{bmatrix} b^{11}(z) & \dots & b^{1m}(z) \\ b^{m1}(z) & \dots & b^{mm}(z) \end{bmatrix} \tag{7.87}$$

where

$$\begin{aligned} a(z) &= z^n + a_1 z^{n-1} + \dots + a_n \\ b^{ij}(z) &= b_m^{ij} z^m + b_{m-1}^{ij} z^{m-1} + \dots + b_0^{ij} \end{aligned} \tag{7.88}$$

It is therefore obvious that equations (7.87) and (7.88) gives

$$
\begin{aligned}
&(I_{n*n}+a_1 z^{-1} I_{n*n}+\ldots+a_n z^{-n} I_{n*n})Y(z)= \\
&\left\{\begin{bmatrix} b_m^{11} & \cdots & b_m^{1m} \\ & b_m^{ij} & \\ b_m^{m1} & \cdots & b_m^{mm} \end{bmatrix} z^{m-n}+\ldots+\begin{bmatrix} b_0^{11} & \cdots & b_0^{1m} \\ & b_0^{ij} & \\ b_0^{m1} & \cdots & b_0^{mm} \end{bmatrix} z^{-n}\right\} U(z)
\end{aligned} \tag{7.89}
$$

where I_{n*n} is the identity matrix and (7.89) can be rewritten as

$$
y_k+\hat{A}_1 y_{k-1}+\ldots+\hat{A}_n y_{k-n}=\hat{B}_{n-m} u_{k-n+m}+\ldots+\hat{B}_n u_{k-n} \tag{7.90}
$$

$\hat{A}_i$ are diagonal matrices and $\hat{B}_i$ are as in equation (7.89). This representation is used in the RLS system identification block and due to the diagonal form of the $\hat{A}_i$ matrices, it can be readily employed in the decentralized control design.

Example 7.6.2

Consider the multivariable plant of example 7.6.1. For the nominal model given by equation (7.84), the set of variable pairings corresponding to $\lambda_{ij}=5$ has a significantly better performance than the other choices. This was shown in example 2.3.2. So, the appropriate input-output pair is $(u_1-y_3, u_2-y_1, u_3-y_2)$. We now consider the parameter variations given in equation (7.85), the resulting RGA matrix is given by equation (7.86). Hence, the parameter changes in the multivariable plant have clearly resulted in a new input-output pairing $(u_1-y_2, u_2-y_3, u_3-y_1,)$.

If a fixed structure decentralized controller is employed, this new input-output pairing can easily lead to an unstable closed-loop plant with the previous control configuration. However, in the case of a reconfigurable control design methodology, the parameters of the RGA matrix are identified as in figure 7.17. Hence, at 1000th second when the change has occurred, a new input-output pairing is chosen. As is shown in figure 7.18, the decentralized reconfigurable controller can easily handle the new structure.

■

Example 7.6.3

Consider the multivariable plants governed by the following state space equation

$$
\begin{aligned}
\dot{x}(t) &= Ax(t)+Bu(t)+Ef(t) \\
y(t) &= Cx(t)
\end{aligned} \tag{7.91}
$$

where $x \in R^n$, $u \in R^m$, $y \in R^m$ and $f(t) \in R^m$ is the time dependent unknown fault vector. Also, A, B, and C are the usual state-space matrices and E is the full rank fault matrix. Equivalently, equation (7.91) can be rewritten as

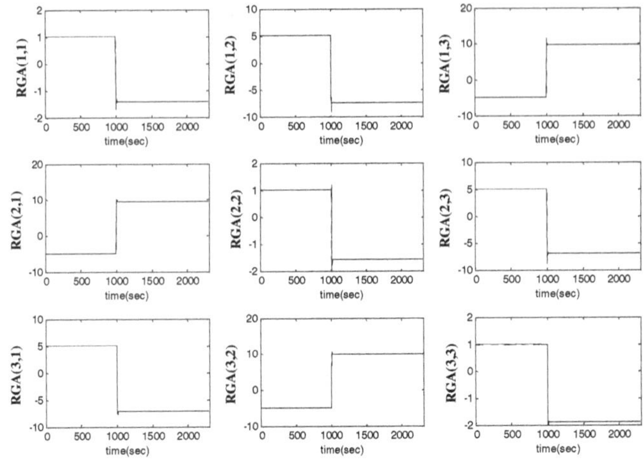

Fig. 7.17 The convergence of the RGA parameters.

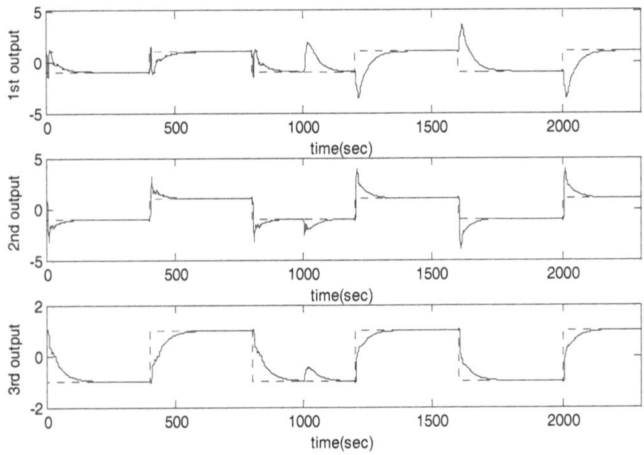

Fig. 7.18 The plant outputs track the reference inputs using the Self Tuning Regulators (STR) and decentralized reconfigurable control structure. Solid lines denote the plant outputs and dashed lines denote the reference inputs.

$$Y(s) = G_p(s)u(s) + G_f(s)f(s) \tag{7.92}$$

where $G_p(s)$ denotes the plant transfer function matrix and $G_f(s)$ is the fault transfer function matrix. Let the pulse transfer function matrix of the multivariable plant be described as

$$G_P(z) = \frac{z - 0.1828}{z^2 - 0.7358z + 0.1353} \begin{bmatrix} 0.3378 & -1.4155 & -8.7697 \\ 2.0911 & 0.3378 & -8.7697 \\ 0.3378 & 0.3378 & 0.3378 \end{bmatrix}$$

and consider the following fault transfer function matrix

$$G_F(z) = \frac{1}{z^2 - 0.7358z + 0.1353} \begin{bmatrix} 4z + 2 & 0 & 0 \\ 4z - 0.8 & 0.8 & 0 \\ 0 & 0 & 4z + 1.2 \end{bmatrix}$$

Also, let the fault vector be

$$f(t) = \begin{cases} 0 & t < 900 \\ [3, 0.05\sin(t), 5]^T & t \geq 900 \end{cases}$$

For $t < 900$, the corresponding RGA is

$$\Lambda_1 = \begin{bmatrix} 1.0009 & 5.0010 & -5.0019 \\ -5.0028 & 1.0009 & 5.0019 \\ 5.0019 & -5.0019 & 1.0000 \end{bmatrix}$$

which similar to example 7.6.2, gives $(u_1 - y_3, u_2 - y_1, u_3 - y_2)$ as the input-output pair. But, for $t \geq 900$ the corresponding RGA converges to

$$\Lambda_2 = \begin{bmatrix} 1.4348 & 0.5311 & -0.9659 \\ -0.8821 & 0.1395 & 1.7426 \\ 0.4473 & 0.3259 & 0.2233 \end{bmatrix}$$

which clearly indicates that the structural changes caused by the fault, lead to the new input-output pair $(u_1 - y_1, u_2 - y_3, u_3 - y_2)$.

Employing the pairing prior to the fault for the case when the fault has occurred will lead to closed-loop instability. To overcome this problem, the decentralized controller is reconfigured according to the new structure. Figure 7.19 show the outputs of the plant.

It is clear that subsequent to the fault and following a short transient after the system reconfiguration, the closed–loop responses are as desired.

■

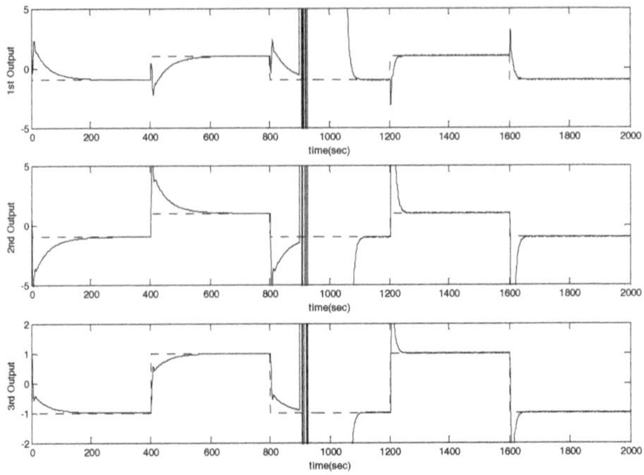

Fig. 7.19 The plant outputs track the reference inputs using the Self Tuning Regulators (STR) and decentralized reconfigurable control structure. Solid lines denote the plant outputs and dashed lines denote the reference inputs.

7.7 Conclusion

This chapter has dealt with the important problem of control configuration for uncertain multivariable plants. By employing the examples and simulation results, it is shown that the input-output pairing of uncertain multivariable plants can change in the face of plant uncertainty. This in turn results in deteriorated or even unstable closed loop behavior of the decentralized control methodology. First, the basic results available on structured uncertainty, input uncertainty and condition number in the multivariable transfer function are presented. These are the pioneer attempts to include the uncertainty concept in input-output pairing analysis. The results are therefore of limited practical interest.

The statistical based approach to uncertainty and control configuration design is the first serious result in the input-output pairing of uncertain multivariable plants. To improve these results the pairing in norm-bounded uncertain plants is considered. This is achieved by developing a signal-based representation of the Relative Gain for uncertain plants. Then, a bound on the magnitude of the worst-case relative gain, calculated at steady state and also at higher frequencies is derived. In addition, the necessary and sufficient conditions for the relative gain sign change in norm-bounded uncertain plants are given.

In this chapter, the upper bounds of the element variations in the DIOPM due to the unstructured uncertainty in the plant are derived and possible changes in input-output pairing resulting from the model uncertainty are detected.

Finally, adaptive input-output pairing and a simple reconfigurable decentralized controller are developed. Adaptive pairing directly identifies the cases that lead to changes in the input-output pair. It also gives an on line pairing selection after the pairing variation. Then, a general reconfigurable decentralized control strategy is presented.

References

Arkun, Y.: Relative sensitivity: a dynamic closed-loop interaction measure and design tool. AIChE Journal 34, 672–675 (1988)

Astrom, K.J., Wittenmark, B.: Adaptive Control. Addison-Wesley, Reading (1995)

Chen, D., Seborg, D.E.: Relative gain array analysis for uncertain process models. AIChE Journal 48, 302–310 (2002)

Fernando, K.V., Nicholson, H.: On the structure of balanced and other principal representations of SISO systems. IEEE Trans. Autom. Contr. 28, 228–231 (1983)

Grosdidier, P., Morari, M., Holt, B.R.: Closed-loop properties from steady state gain information. Ind. Eng. Chem. Fundam. 24, 221–235 (1985)

Haggblom, K.E.: Limitations and use of the RGA as a controllability measure. In: Proceeding of the Automation Days, Helsinki, pp. 178–183 (1995)

Hovd, M., Skogestad, S.: Simple frequency-dependent tools for control system analysis, structure selection and design. Automatica 28, 989–996 (1992)

Kariwala, V., Skogestad, S., Forbes, J.F.: Relative gain array for norm-bounded uncertain systems. Ind. Eng. Chem. Res. 45, 1751–1757 (2006)

Kariwala, V., Hovd, M.: Relative Gain Array: Common misconceptions and clarifications. In: Proceedings of the 7th Symposium on Computer Process Control, Lake Louise, Canada (2006)

Khaki-Sedigh, A., Moaveni, B.: Relative gain array Analysis of uncertain multivariable plants. In: Proceeding of the 7th European Control Conference, Cambridge, UK (2003a)

Khaki-Sedigh, A., Moaveni, B.: Adaptive input-output pairing using online RGA identification. In: Proceeding of the 1st African Control Conference, Cape Town, South Africa (2003b)

Moaveni, B., Khaki-Sedigh, A.: Further theoretical results on relative gain array for norm-bounded uncertain systems. Ind. Eng. Chem. Res. 46, 8288–8289 (2007a)

Moaveni, B., Khaki-Sedigh, A.: Reconfigurable Controller Design for Linear Multivariable Systems. Int. J. Modeling, Identification and Control 2, 138–146 (2007b)

Moaveni, B., Khaki-Sedigh, A.: Input-output pairing analysis for uncertain multivariable processes. J. Process Contr. 18, 527–532 (2008)

Nett, C.N., Manousiouthakis, V.: Euclidean condition and block relative gain: connections, conjectures, and clarifications. IEEE Trans. Autom. Control 32, 405–407 (1987)

Ogunnaike, B.A., Pay, W.H.: Processes, dynamics, modeling and control. Oxford University Press, Oxford (1994)

Skogestad, S., Morari, M.: Implications of large RGA elements on control performance. Ind. Eng. Chem. Res. 26, 2323–2330 (1987)

Skogestad, S., Hovd, M.: Use of frequency-dependent RGA for control structure selection. In: Proceeding of the American Control Conference, pages, pp. 2133–2139 (1990)

Skogestad, S., Postlethwaite, I.: Multivariable feedback control analysis and design. Wiley, Chichester (2005)

Yu, C.C., Luyben, W.L.: Robustness with respect to integral controllability. Ind. Eng. Chem. Res. 26, 1043–1045 (1987)

Zhu, Z.X., Jutan, A.: A new variable pairing criterion based on Niederlinski index. Chem. Eng. Comm. 121, 235–250 (1993)

Appendix: Mathematical Models Used in Examples

A.1 Distillation Column Transfer Function Matrices

A.1.1 Wood and Berry distillation column transfer function matrix (Wood and Berry 1973)

$$\begin{bmatrix} X_D(s) \\ X_B(s) \end{bmatrix} = \begin{bmatrix} \dfrac{12.8e^{-s}}{16.7s+1} & \dfrac{-18.9e^{-s}}{21s+1} \\ \dfrac{6.6e^{-7s}}{10.9s+1} & \dfrac{-19.4e^{-s}}{14.4s+1} \end{bmatrix} \begin{bmatrix} R(s) \\ S(s) \end{bmatrix}$$

and its state space model (using a 1st order Pade approximation for the delays) is

$$A = \begin{bmatrix} -2.4373 & -0.9235 & -0.0992 & 0.0024 & -0.0007 & 0.0011 & -0.0016 \\ 1 & 0 & 0 & 0 & 0 & 0 & 0 \\ 0 & 1 & 0 & 0 & 0 & 0 & 0 \\ 0 & 0 & -0.7603 & -0.0003 & -0.5054 & 0.2820 & 0.4122 \\ 0 & 0 & 0.2253 & -0.0014 & -1.3007 & -0.6874 & 0.1177 \\ 0 & 0 & -0.3379 & -0.0001 & 0.7754 & 0.1253 & -0.2612 \\ 0 & 0 & 0.5069 & 0.0002 & 0.3369 & 0.8120 & 0.3919 \end{bmatrix}$$

$$B^T = \begin{bmatrix} 1 & 0 & 0 & 0 & 0 & 0 & 0 \\ 0 & 0 & 0 & 0.1713 & 0.9492 & 0.0761 & -0.1142 \end{bmatrix}$$

$$C = \begin{bmatrix} -0.7665 & 1.2436 & 0.5585 & -0.0583 & 0.9091 & 0.0489 & -0.3796 \\ -0.6055 & -1.0743 & 0.2838 & -0.0443 & 1.3519 & 0.0572 & -0.5883 \end{bmatrix}$$

The inputs are the reflux flow rate and the steam flow rate to the reboiler, and the outputs are the overhead and bottoms compositions of methanol, respectively.

A.1.2 The transfer function matrix of a distillation column plant (Ogunnaike and Pay 1994)

$$G(s)=\begin{bmatrix} \dfrac{-21.6e^{-s}}{8.5s+1} & \dfrac{1.26e^{-0.3s}}{7.05s+1} \\ \dfrac{-2.75e^{-1.8s}}{8.2s+1} & \dfrac{-4.28e^{-0.35s}}{9.0s+1} \end{bmatrix}$$

A.1.3 The transfer function matrix of Doukas and Luyben distillation column (Grosdidier and Morari 1986)

$$G(s)=\begin{bmatrix} \dfrac{0.374e^{-7.75s}}{(22.2s+1)^2} & \dfrac{-11.3e^{-3.79s}}{(21.74s+1)^2} & \dfrac{-9.811e^{-1.59s}}{(11.36s+1)} \\ \dfrac{-1.986e^{-0.71s}}{(66.67s+1)^2} & \dfrac{5.24e^{-60s}}{(400s+1)} & \dfrac{5.984e^{-2.24s}}{(14.29s+1)} \\ \dfrac{0.0204e^{-0.59s}}{(7.14s+1)^2} & \dfrac{-0.33e^{-0.68s}}{(2.38s+1)^2} & \dfrac{2.38e^{-0.42s}}{(1.43s+1)^2} \end{bmatrix}$$

A.1.4 The transfer function matrix of a distillation column (Bao et al. 2007)

$$G(s)=\begin{bmatrix} \dfrac{-1.986e^{-0.71s}}{66.67s+1} & \dfrac{5.24e^{-60s}}{400s+1} & \dfrac{5.984e^{-2.24s}}{14.29s+1} \\ \dfrac{0.0204e^{-4.199s}}{5s+1} & \dfrac{-0.33e^{-1.883s}}{3.904s+1} & \dfrac{2.38e^{-1.143s}}{10s+1} \\ \dfrac{0.374e^{-7.75s}}{22.22s+1} & \dfrac{-11.3e^{-14.78s}}{35.66s+1} & \dfrac{-9.881e^{-1.59s}}{11.35s+1} \end{bmatrix}$$

A.1.5 The transfer function matrix of Alatiqi and Luyben column/stripper distillation plant (Kariwala et al. 2003)

$$G(s)=\begin{bmatrix} \dfrac{4.09e^{-1.3s}}{(33s+1)(8.3s+1)} & \dfrac{-6.36e^{-1.2s}}{(31.6s+1)(20s+1)} & \dfrac{-0.25e^{-1.4s}}{(21s+1)} & \dfrac{-0.49e^{-6s}}{(22s+1)^2} \\ \dfrac{-4.17e^{-5s}}{(45s+1)} & \dfrac{6.93e^{-1.02s}}{(44.6s+1)} & \dfrac{-0.05e^{-6s}}{(34.5s+1)^2} & \dfrac{1.53e^{-3.8s}}{(48s+1)} \\ \dfrac{1.73e^{-18s}}{(13s+1)^2} & \dfrac{5.11e^{-12s}}{(13.3s+1)^2} & \dfrac{4.61e^{-1.01s}}{(18.5s+1)} & \dfrac{-5.49e^{-1.5s}}{(15s+1)} \\ \dfrac{-11.2e^{-2.6s}}{(43s+1)(6.5s+1)} & \dfrac{14(10s+1)e^{-0.02s}}{(45s+1)(17.4s^2+3s+1)} & \dfrac{0.1e^{-0.05s}}{(31.6s+1)(5s+1)} & \dfrac{4.49e^{-0.6s}}{(48s+1)(6.3s+1)} \end{bmatrix}$$

A.1.6 The transfer function matrix of a side stream distillation column (Doukas and Luyben 1978)

$$G(s)=\begin{bmatrix} \dfrac{-9.811e^{-1.59s}}{11.36s+1} & \dfrac{0.374e^{-7.75s}}{(22.2s+1)^2} & \dfrac{-11.3e^{-3.79s}}{(21.74s+1)^2} \\ \dfrac{5.984e^{-2.24s}}{14.29s+1} & \dfrac{-1.986e^{-0.71s}}{(66.67s+1)^2} & \dfrac{5.24e^{-60s}}{(400s+1)^2} \\ \dfrac{2.38e^{-0.42s}}{(1.43s+1)^2} & \dfrac{0.0204e^{-0.59s}}{(7.14s+1)^2} & \dfrac{-0.33e^{-0.68s}}{(2.38s+1)^2} \\ \dfrac{-11.67e^{-1.91s}}{12.19s+1} & \dfrac{-0.176e^{-0.48s}}{(6.9s+1)^2} & \dfrac{4.48e^{-0.52s}}{(11.11s+1)^2} \end{bmatrix}$$

A.2 State Space Models

A.2.1 The state space model of a three-stage distillation column (Shimizu and Matsubara 1985)

$$\dot{x}=\begin{bmatrix} -14 & 12 & 0 \\ 23 & -51 & 35 \\ 0 & 5 & -11 \end{bmatrix}x+\begin{bmatrix} 0 \\ -10 \\ -9 \end{bmatrix}p+\begin{bmatrix} 0 & 0 & 0 \\ 3.8 & -2.9 & 0 \\ 0.59 & -2.9 & 0 \end{bmatrix}u$$

$$\varepsilon\dot{p}=\begin{bmatrix} 0 & 10 & 3 \end{bmatrix}x+\begin{bmatrix} -4 \end{bmatrix}p+\begin{bmatrix} -0.2 & -2.2 & 3.3 \end{bmatrix}u$$

$$y=\begin{bmatrix} 1 & 0 & 0 \\ 0 & 0 & 1 \\ 0 & 0 & 0 \end{bmatrix}x+\begin{bmatrix} 0 \\ 0 \\ 1 \end{bmatrix}p$$

This model can be used for singular perturbation type studies, where $\varepsilon \to 0$ is the perturbation parameter and results in fast and slow decomposition of the plant.

A.2.2 The Quadruple-tank plant (Johansson 2000)

$$\dot{h}_1=-\frac{a_1}{A_1}\sqrt{2gh_1}+\frac{a_3}{A_1}\sqrt{2gh_3}+\frac{\gamma_1 k_1}{A_1}v_1$$

$$\dot{h}_2=-\frac{a_2}{A_2}\sqrt{2gh_2}+\frac{a_4}{A_2}\sqrt{2gh_4}+\frac{\gamma_2 k_2}{A_2}v_2$$

$$\dot{h}_3=-\frac{a_3}{A_3}\sqrt{2gh_3}+\frac{(1-\gamma_2)k_2}{A_3}v_2$$

$$\dot{h}_4=-\frac{a_4}{A_4}\sqrt{2gh_4}+\frac{(1-\gamma_1)k_1}{A_4}v_1$$

where h_i, A_i and a_i for $i = 1,\ldots,4$ denote the level, cross-section and outlet hole cross sections of the ith tank, respectively. Also, g denotes the gravity acceleration. In the above equations, $k_i v_i$ is the corresponding flow to the pump voltage, v_i. Moreover, $\gamma_1, \gamma_2 \in (0,1)$ indicate the division ratio of flow using the valves. The control objective is the control of two lower tanks levels, h_1 and h_2, using the two pumps.

The transfer function model of the quadruple tank is

$$G(s) = \begin{bmatrix} \dfrac{\gamma_1 T_1 k_1 k_c}{A_1(sT_1+1)} & \dfrac{(1-\gamma_2)T_1 k_2 k_c}{A_1(sT_1+1)(sT_3+1)} \\ \dfrac{(1-\gamma_1)T_2 k_1 k_c}{A_2(sT_2+1)(sT_4+1)} & \dfrac{\gamma_2 T_2 k_2 k_c}{A_2(sT_2+1)} \end{bmatrix}$$

where

$$T_i = \frac{A_i}{a_i}\sqrt{\frac{2h_i^0}{g}}$$

and h_i^0 are the initial levels.

This multivariable plant exhibit different behaviors for variations in γ_1 and γ_2. The plant can be minimum or non-minimum phase with different input-output pairs.

A.2.3 The nonlinear state space model for a cascade of two continuous stirred tank reactor (CSTR) (Daoutidis and Kravaris 1992).

$$\begin{aligned}
\frac{dc_{A1}}{dt} &= \tfrac{F}{V}(c_{A0} - c_{A1}) - k_0 e^{-E/RT_1} c_{A1}^2 \\
\frac{dc_{A2}}{dt} &= \tfrac{F}{V}(c_{A1} - c_{A2}) - k_0 e^{-E/RT_2} c_{A2}^2 \\
\frac{dT_1}{dt} &= \tfrac{F}{V}(T_0 - T_1) + \tfrac{-\Delta H}{\rho c_p} k_0 e^{-E/RT_1} c_{A1}^2 + \tfrac{1}{V\rho c_p} Q_1 \\
\frac{dT_2}{dt} &= \tfrac{F}{V}(T_1 - T_2) + \tfrac{-\Delta H}{\rho c_p} k_0 e^{-E/RT_2} c_{A2}^2 + \tfrac{1}{V\rho c_p} Q_2
\end{aligned}$$

where c_p is the heat capacity, E is the activation energy, F is the volumetric flow rate, Q_1, Q_2 are the heat inputs to tanks 1 and 2, T_1, T_2 are the temperatures in tanks 1 and 2, T_0 is the inlet temperature, V is the volume, $-\Delta H$ is the heat of reaction, c_{A1}, c_{A2} are the molar concentration of A in the tanks, c_{A0} is the inlet molar concentration of A, k_0 is the Arrhenius frequency factor, ρ is the density, $y = c_{A2}$ is the plant output, and $u_1 = Q_1$, $u_2 = Q_2$ are the process inputs.

A.2.4 The nonlinear state space model of a single-effect evaporator (Daoutidis and Kravaris 1992)

$$\dot{x} = \begin{bmatrix} 0 \\ -\dfrac{B_s x_2}{Ac(x_1 + h_s)} \end{bmatrix} + \begin{bmatrix} -\dfrac{1}{Ac} & -\dfrac{1}{\Delta H_v Ac} \\ 0 & \dfrac{x_2 + x_{Bs}}{\Delta H_v Ac(x_1 + h_s)} \end{bmatrix} \begin{bmatrix} u_1 \\ u_2 \end{bmatrix} + \begin{bmatrix} 0 \\ \dfrac{F}{Ac(x_1 + h_s)} \end{bmatrix} d$$

$$y = \begin{bmatrix} y_1 \\ y_2 \end{bmatrix} = \begin{bmatrix} x_1 \\ x_2 \end{bmatrix}$$

$x_1 = h - h_s$, $x_2 = x_B - x_{Bs}$, $u_1 = B - B_s$, $u_2 = Q - Q_s$, $d = x_F - x_{Fs}$ the subscript s denotes the nominal steady-state value and d is the disturbance input. Also, A denotes the cross-section area, F, B and D indicate the molar flow rates and in the steady-state $F_s = B_s + D_s$, c denotes the molar density of the feed and bottom streams, h is the liquid level in the evaporator and x_F, x_B indicate the solute concentration at the feed and bottom streams, respectively.

A.2.5 The simplified state space model of a free gyroscope

$$\dot{x}_1 = x_2$$

$$\dot{x}_2 = \frac{1}{I_r \cos x_3}(T_z + I_R \omega_s x_4)$$

$$\dot{x}_3 = x_4$$

$$\dot{x}_4 = \frac{1}{I_r}(T_y - I_R \omega_s x_2 \cos x_3)$$

where $x_1 = \psi_D, x_2 = \dot{\psi}_D, x_3 = \theta_D, x_2 = \dot{\theta}_D$. Plant inputs are the torques $u_1 = T_z$ and $u_2 = T_y$, the outputs are $y_1 = \psi_D$ and $y_2 = \theta_D$, yaw and pitch angels of the rotor respectively. Also, I_r and I_R denote the moment of inertia in pitch and yaw respectively and ω_s shows the rotor angular speed.

A.2.6 The nonlinear state space model for a cascade of two continuous stirred tank reactor (CSTR) (Daoutidis and Kravaris 1992)

$$\dot{x}_1 = u_1 + u_2 - u_3$$

$$\dot{x}_2 = \frac{F_{As}(c_{A0} - x_2 - c_{As}) - F_{Bs}(x_2 + c_{As})}{x_1 + V_s} - k(x_2 + c_{As})(x_3 + c_{Bs})e^{-E/R(x_5+T_s)}$$
$$+ \frac{u_1(c_{A0} - x_2 - c_{As}) - u_2(x_2 + c_{As})}{x_1 + V_s}$$

$$\dot{x}_3 = \frac{F_{Bs}(c_{B0} - x_3 - c_{Bs}) - F_{As}(x_3 + c_{Bs})}{x_1 + V_s} - k(x_2 + c_{As})(x_3 + c_{Bs})e^{-E/R(x_5+T_s)}$$
$$+ \frac{u_2(c_{B0} - x_3 - c_{Bs}) - u_1(x_3 + c_{Bs})}{x_1 + V_s}$$

$$\dot{x}_4 = -\frac{(F_{As} + F_{Bs})(x_4 + c_{cs})}{x_1 + V_s} + k(x_2 + c_{As})(x_3 + c_{Bs})e^{-E/R(x_5+T_s)} - \frac{(x_4 + c_{cs})(u_1 + u_2)}{x_1 + V_s}$$

$$\dot{x}_5 = \frac{(T_A - x_5 - T_s)F_{As} + (T_B - x_5 - T_s)F_{Bs}}{x_1 + V_s} + \frac{Q_s}{\rho c_p(x_1 + V_s)}$$
$$- \frac{k\Delta H}{\rho c_p}(x_2 + c_{As})(x_3 + c_{Bs})e^{-E/R(x_5+T_s)}$$
$$+ \frac{(T_A - x_5 - T_s)u_1 + (T_B - x_5 - T_s)u_2}{x_1 + V_s} + \frac{u_4}{\rho c_p(x_1 + V_s)}$$

In this process, two solution streams consisting of species *A* and *B*, at volumetric flow rates F_A and F_B, temperatures T_A and T_B, and concentrations c_{A0} and c_{B0}, enter the reactor. The effluent stream leaves the reactor at a flow rate F, concentrations c_A, c_B, c_C, c_D, and temperature T. Heat may be added or removed from the system at a rate and volume Q and V using an appropriate heating/cooling system. Also, assuming constant density and heat capacity ρ and c_p for the liquid streams and neglecting the heat of solution effects, the material and energy balances equations give a nonlinear state space model of the process. Also

$$x_1 = V - V_s,\ x_2 = c_A - c_{As},\ x_3 = c_B - c_{Bs},\ x_4 = c_C - c_{Cs},\ x_5 = T - T_s$$

and

$$u_1 = F_A - F_{As},\ u_2 = F_B - F_{Bs},\ u_3 = F - F_s,\ u_4 = Q - Q_s$$

and

$$y_1 = x_1,\ y_2 = x_2,\ y_3 = x_4,\ y_4 = x_5$$

where the subscript "s" denotes a nominal steady state value.

A.3 Nonphysical Transfer Function Matrices

Some transfer function matrices of linear multivariable plants introduced in the literature as bench mark models are given below.

A.3.1 (Mc Avoy et al. 2003)

$$G(s)=\begin{bmatrix} \dfrac{5e^{-40s}}{100s+1} & \dfrac{e^{-4s}}{10s+1} \\ \dfrac{-5e^{-4s}}{10s+1} & \dfrac{5e^{-40s}}{100s+1} \end{bmatrix}$$

A.3.2 (Hovd and Skogestad 1994)

$$G(s)=\begin{bmatrix} \dfrac{9s+1}{(-s+1)(s+1)} & \dfrac{2s-18}{(-s+1)(s+1)} \\ \dfrac{-1.5s-6}{(-s+1)(0.5s+1)} & \dfrac{12}{(-s+1)(0.5s+1)} \end{bmatrix}$$

A.3.3 (McAvoy 1983)

$$G(s)=\begin{bmatrix} \dfrac{0.562}{(7.74s+1)^2} & \dfrac{-0.516e^{-0.5s}}{(7.1s+1)^2} \\ \dfrac{0.344e^{-0.5s}}{(15.8s+1)(0.5s+1)} & \dfrac{-0.394}{(13.8s+1)(0.4s+1)} \end{bmatrix}$$

A.3.4 (McAvoy 1983)

$$G(s)=\begin{bmatrix} \dfrac{-0.805}{(18.3s+1)(5.6s+1)} & \dfrac{0.055}{(5.76s+1)(1.25s+1)} \\ \dfrac{-0.465e^{-0.3s}}{(28.3s+1)(0.62s+1)} & \dfrac{-0.055}{(3.3s+1)} \end{bmatrix}$$

A.3.5 (Salgado and Conley 2004)

$$G(z)=\begin{bmatrix} \dfrac{0.5}{z-0.5} & \dfrac{0.15}{(z-0.8)z^{l}} \\ \dfrac{0.1}{(z-0.5)(z-0.8)} & \dfrac{0.3}{z-0.7} \end{bmatrix}$$

A.3.6 (Wittemark and Salgado 2002)

$$G(z)=\begin{bmatrix} \dfrac{0.1021}{z-0.9048} & \dfrac{0.3707z-0.3535}{z^2-1.724z+0.7408} \\ \dfrac{-0.192z+0.1826}{z^2-1.869z+0.8781} & \dfrac{0.09516}{z-0.9048} \end{bmatrix}$$

A.3.7 (Moaveni and Khaki Sedigh 2008)

$$G(s)=\begin{bmatrix} \dfrac{-0.9019s+15.47}{s^2+9.163s+15.47} & \dfrac{-3.327}{s+6.931} \\ \dfrac{0.8926}{s+2.231} & \dfrac{0.7549s+13.92}{s^2+9.163s+15.47} \end{bmatrix}$$

A.3.8 (Grosdidier and Morari 1986)

$$G(s)=\begin{bmatrix} \dfrac{5}{4s+1} & \dfrac{2.5e^{-5s}}{(2s+1)(15s+1)} \\ \dfrac{-4e^{-6s}}{20s+1} & \dfrac{1}{3s+1} \end{bmatrix}$$

A.3.9 (Chen and Seborg 2002)

$$G(s)=\begin{bmatrix} \dfrac{0.66e^{-2.6s}}{6.7s+1} & \dfrac{-0.61e^{-3.5s}}{8.64s+1} & \dfrac{-0.0049e^{-s}}{9.06s+1} \\ \dfrac{1.11e^{-0.65s}}{3.25s+1} & \dfrac{-2.36e^{-3s}}{5s+1} & \dfrac{-0.012e^{-1.2s}}{7.09s+1} \\ \dfrac{-33.68e^{-9.2s}}{8.15s+1} & \dfrac{46.2e^{-9.4s}}{10.9s+1} & \dfrac{0.87(11.61s+1)e^{-s}}{(3.89s+1)(18.8s+1)} \end{bmatrix}$$

A.3.10 (Hovd and Skogestad 1992)

$$G(s)=\frac{1-s}{(5s+1)^2}\begin{bmatrix} 1 & -4.19 & -25.96 \\ 6.19 & 1 & -25.96 \\ 1 & 1 & 1 \end{bmatrix}$$

A.3.11 (Huang et al. 1994)

$$G(s)=\begin{bmatrix}\dfrac{-2e^{-s}}{10s+1} & \dfrac{1.5e^{-s}}{s+1} & \dfrac{e^{-s}}{s+1}\\ \dfrac{1.5e^{-s}}{s+1} & \dfrac{e^{-s}}{s+1} & \dfrac{-2e^{-s}}{10s+1}\\ \dfrac{e^{-s}}{s+1} & \dfrac{-2e^{-s}}{10s+1} & \dfrac{1.5e^{-s}}{s+1}\end{bmatrix}$$

A.3.12 (Mc Avoy et al. 2003)

$$G(s)=\begin{bmatrix}\dfrac{5e^{-40s}}{100s+1} & \dfrac{k_{12}e^{-4s}}{10s+1}\\ \dfrac{k_{21}e^{-4s}}{10s+1} & \dfrac{5e^{-40s}}{100s+1}\end{bmatrix}$$

with values of $k_{12}=1$, and $k_{21}=-5,\ 1,$ and 5.

A.3.13 (Huang et al. 1994)

$$G(s)=\begin{bmatrix}\dfrac{-2e^{-s}}{10s+1} & \dfrac{1.5e^{-s}}{s+1}\\ \dfrac{1.5e^{-s}}{s+1} & \dfrac{-2e^{-s}}{10s+1}\end{bmatrix}$$

A.3.14 (Salgado and Conley 2004)

$$G(s)=\begin{bmatrix}\dfrac{0.4}{(s+1)^2} & \dfrac{4(s+3)}{(s+2)(s+5)} & \dfrac{-2}{s+4}\\ \dfrac{2}{(s+2)(s+1)} & \dfrac{2}{(s+2)^2} & \dfrac{1}{s+2}\\ \dfrac{6(-s+1)}{(s+5)(s+4)} & \dfrac{4}{(s+3)^2} & \dfrac{8}{(s+2)(s+5)}\end{bmatrix}$$

References

Bao, J., Chan, K.H., Zhang, W.Z., Lee, P.L.: An experimental pairing method for multiloop control based on passivity. J. Process Contr. 17, 787–798 (2007)

Chang, J.W., Yu, C.C.: The relative gain for non-square multivariable systems. Chem. Eng. Sci. 45, 1309–1323 (1990)

Chen, D., Seborg, D.E.: Relative gain array analysis for uncertain process models. AIChE Journal 48, 302–310 (2002)

Daoutidis, P., Kravaris, C.: Structural evaluation of control configuration for multivariable nonlinear processes. Chem. Eng. Sci. 47, 1091–1107 (1992)

Grosdidier, P., Morari, M.: Interaction measures for systems under decentralized control. Automatica 22, 309–319 (1986)

Huang, H.P., Ohshima, M., Hashimoto, I.: Dynamic interaction and multiloop control system design. J. Process Contr. 4, 15–27 (1994)

Hovd, M., Skogestad, S.: Simple frequency-dependent tools for control system analysis, structure selection and design. Automatica 28, 989–996 (1992)

Hovd, M., Skogestad, S.: Pairing criteria for decentralized control of unstable plants. Ind. Eng. Chem. Res. 33, 2134–2139 (1994)

Huang, H.P., Ohshima, M., Hashimoto, I.: Dynamic interaction and multiloop control system design. J. Process Contr. 4, 15–27 (1994)

Johansson, K.H.: The Quadruple-tank process: a multivariable laboratory process with an adjustable zero. IEEE Trans. Control Syst. Technol. 8, 456–465 (2000)

Kariwala, V., Forbes, J.F., Meadows, E.S.: Block relative gain: properties, and pairing rules. Ind. Eng. Chem. Res. 42, 4564–4574 (2003)

McAvoy, T.J.: Some results on dynamic interaction analysis of complex control systems. Ind. Eng. Chem. Des. Dev. 22, 42–49 (1983)

McAvoy, T., Arkun, Y., Chen, R., Robinson, D., Schnelle, P.D.: A new approach to defining a dynamic relative gain. Control Eng. Pract. 11, 907–914 (2003)

Moaveni, B., Khaki-Sedigh, A.: A new approach to compute the cross-Gramian matrix and its application in input-output pairing of linear multivariable plants. J. Applied Sciences 8, 608–614 (2008)

Ogunnaike, B.A., Pay, W.H.: Processes, dynamics, modeling and control. Oxford University Press, Oxford (1994)

Salgado, M.E., Conley, A.: MIMO interaction measure and controller structure selection. Int. J. Control 77, 367–383 (2004)

Shimizu, K., Matsubara, M.: Singular perturbation for the dynamic interaction measure. IEEE Trans. Autom. Control 30, 790–792 (1985)

Wittenmark, B., Salgado, M.E.: Hankel-norm based interaction measure for input-output pairing. In: Proceeding of the 15th IFAC World Congress on Automatic Control, Barcelona, Spain (2002)

Wood, R.K., Berry, M.W.: Terminal composition control of a binary distillation column. Chem. Eng. Sci. 28, 1707–1717 (1973)

Index

Lecture Notes in Control and Information Sciences

Edited by M. Thoma, F. Allgöwer, M. Morari

Further volumes of this series can be found on our homepage:
springer.com

Vol. 391: Khaki-Sedigh, A.; Moaveni, B.:
Control Configuration Selection for Multivariable Plants
232 p. [978-3-642-03192-2]

Vol. 390: Chesi, G.; Garulli, A.;
Tesi, A.; Vicino, A.:
Homogeneous Polynomial Forms for Robustness Analysis of Uncertain Systems
197 p. [978-1-84882-780-6]

Vol. 389: Bru, R.; Romero-Vivó, S. (Eds.):
Positive Systems
398 p. [978-3-642-02893-9]

Vol. 388: Jacques Loiseau, J.; Michiels, W.;
Niculescu, S-I.; Sipahi, R. (Eds.):
Topics in Time Delay Systems
418 p. [978-3-642-02896-0]

Vol. 387: Xia, Y.;
Fu, M.; Shi, P.:
Analysis and Synthesis of Dynamical Systems with Time-Delays
283 p. 2009 [978-3-642-02695-9]

Vol. 386: Huang, D.;
Nguang, S.K.:
Robust Control for Uncertain Networked Control Systems with Random Delays
159 p. 2009 [978-1-84882-677-9]

Vol. 385: Jungers, R.:
The Joint Spectral Radius
144 p. 2009 [978-3-540-95979-3]

Vol. 384: Magni, L.; Raimondo, D.M.;
Allgöwer, F. (Eds.):
Nonlinear Model Predictive Control
572 p. 2009 [978-3-642-01093-4]

Vol. 383: Sobhani-Tehrani E.;
Khorasani K.;
Fault Diagnosis of Nonlinear Systems Using a Hybrid Approach
360 p. 2009 [978-0-387-92906-4]

Vol. 382: Bartoszewicz A.;
Nowacka-Leverton A.;
Time-Varying Sliding Modes for Second and Third Order Systems
192 p. 2009 [978-3-540-92216-2]

Vol. 381: Hirsch M.J.; Commander C.W.;
Pardalos P.M.; Murphey R. (Eds.)
Optimization and Cooperative Control Strategies: Proceedings of the 8th International Conference on Cooperative Control and Optimization
459 p. 2009 [978-3-540-88062-2]

Vol. 380: Basin M.
New Trends in Optimal Filtering and Control for Polynomial and Time-Delay Systems
206 p. 2008 [978-3-540-70802-5]

Vol. 379: Mellodge P.; Kachroo P.;
Model Abstraction in Dynamical Systems: Application to Mobile Robot Control
116 p. 2008 [978-3-540-70792-9]

Vol. 378: Femat R.; Solis-Perales G.;
Robust Synchronization of Chaotic Systems Via Feedback
199 p. 2008 [978-3-540-69306-2]

Vol. 377: Patan K.
Artificial Neural Networks for the Modelling and Fault Diagnosis of Technical Processes
206 p. 2008 [978-3-540-79871-2]

Vol. 376: Hasegawa Y.
Approximate and Noisy Realization of Discrete-Time Dynamical Systems
245 p. 2008 [978-3-540-79433-2]

Vol. 375: Bartolini G.; Fridman L.; Pisano A.;
Usai E. (Eds.)
Modern Sliding Mode Control Theory
465 p. 2008 [978-3-540-79015-0]

Vol. 374: Huang B.; Kadali R.
Dynamic Modeling, Predictive Control and Performance Monitoring
240 p. 2008 [978-1-84800-232-6]

Vol. 373: Wang Q.-G.; Ye Z.; Cai W.-J.;
Hang C.-C.
PID Control for Multivariable Processes
264 p. 2008 [978-3-540-78481-4]

Vol. 372: Zhou J.; Wen C.
Adaptive Backstepping Control of Uncertain Systems
241 p. 2008 [978-3-540-77806-6]

Vol. 371: Blondel V.D.; Boyd S.P.; Kimura H. (Eds.)
Recent Advances in Learning and Control
279 p. 2008 [978-1-84800-154-1]

Vol. 370: Lee S.; Suh I.H.; Kim M.S. (Eds.)
Recent Progress in Robotics: Viable Robotic Service to Human
410 p. 2008 [978-3-540-76728-2]

Vol. 369: Hirsch M.J.; Pardalos P.M.; Murphey R.; Grundel D.
Advances in Cooperative Control and Optimization
423 p. 2007 [978-3-540-74354-5]

Vol. 368: Chee F.; Fernando T.
Closed-Loop Control of Blood Glucose
157 p. 2007 [978-3-540-74030-8]

Vol. 367: Turner M.C.; Bates D.G. (Eds.)
Mathematical Methods for Robust and Nonlinear Control
444 p. 2007 [978-1-84800-024-7]

Vol. 366: Bullo F.; Fujimoto K. (Eds.)
Lagrangian and Hamiltonian Methods for Nonlinear Control 2006
398 p. 2007 [978-3-540-73889-3]

Vol. 365: Bates D.; Hagström M. (Eds.)
Nonlinear Analysis and Synthesis Techniques for Aircraft Control
360 p. 2007 [978-3-540-73718-6]

Vol. 364: Chiuso A.; Ferrante A.; Pinzoni S. (Eds.)
Modeling, Estimation and Control
356 p. 2007 [978-3-540-73569-4]

Vol. 363: Besançon G. (Ed.)
Nonlinear Observers and Applications
224 p. 2007 [978-3-540-73502-1]

Vol. 362: Tarn T.-J.; Chen S.-B.; Zhou C. (Eds.)
Robotic Welding, Intelligence and Automation
562 p. 2007 [978-3-540-73373-7]

Vol. 361: Méndez-Acosta H.O.; Femat R.; González-Álvarez V. (Eds.):
Selected Topics in Dynamics and Control of Chemical and Biological Processes
320 p. 2007 [978-3-540-73187-0]

Vol. 360: Kozlowski K. (Ed.)
Robot Motion and Control 2007
452 p. 2007 [978-1-84628-973-6]

Vol. 359: Christophersen F.J.
Optimal Control of Constrained Piecewise Affine Systems
190 p. 2007 [978-3-540-72700-2]

Vol. 358: Findeisen R.; Allgöwer F.; Biegler L.T. (Eds.): Assessment and Future Directions of Nonlinear Model Predictive Control
642 p. 2007 [978-3-540-72698-2]

Vol. 357: Queinnec I.; Tarbouriech S.; Garcia G.; Niculescu S.-I. (Eds.):
Biology and Control Theory: Current Challenges
589 p. 2007 [978-3-540-71987-8]

Vol. 356: Karatkevich A.:
Dynamic Analysis of Petri Net-Based Discrete Systems
166 p. 2007 [978-3-540-71464-4]

Vol. 355: Zhang H.; Xie L.:
Control and Estimation of Systems with Input/Output Delays
213 p. 2007 [978-3-540-71118-6]

Vol. 354: Witczak M.:
Modelling and Estimation Strategies for Fault Diagnosis of Non-Linear Systems
215 p. 2007 [978-3-540-71114-8]

Vol. 353: Bonivento C.; Isidori A.; Marconi L.; Rossi C. (Eds.)
Advances in Control Theory and Applications
305 p. 2007 [978-3-540-70700-4]

Vol. 352: Chiasson, J.; Loiseau, J.J. (Eds.)
Applications of Time Delay Systems
358 p. 2007 [978-3-540-49555-0]

Vol. 351: Lin, C.; Wang, Q.-G.; Lee, T.H., He, Y.
LMI Approach to Analysis and Control of Takagi-Sugeno Fuzzy Systems with Time Delay
204 p. 2007 [978-3-540-49552-9]

Vol. 350: Bandyopadhyay, B.; Manjunath, T.C.; Umapathy, M.
Modeling, Control and Implementation of Smart Structures 250 p. 2007 [978-3-540-48393-9]

Vol. 349: Rogers, E.T.A.; Galkowski, K.; Owens, D.H.
Control Systems Theory and Applications for Linear Repetitive Processes
482 p. 2007 [978-3-540-42663-9]

Vol. 347: Assawinchaichote, W.; Nguang, K.S.; Shi P.
Fuzzy Control and Filter Design for Uncertain Fuzzy Systems
188 p. 2006 [978-3-540-37011-6]

Vol. 346: Tarbouriech, S.; Garcia, G.; Glattfelder, A.H. (Eds.)
Advanced Strategies in Control Systems with Input and Output Constraints
480 p. 2006 [978-3-540-37009-3]

Vol. 345: Huang, D.-S.; Li, K.; Irwin, G.W. (Eds.)
Intelligent Computing in Signal Processing and Pattern Recognition
1179 p. 2006 [978-3-540-37257-8]

Vol. 344: Huang, D.-S.; Li, K.; Irwin, G.W. (Eds.)
Intelligent Control and Automation
1121 p. 2006 [978-3-540-37255-4]

Kolibri 360 GmbH
Ferdinand-Jühlke-Straße 7
99095 Erfurt, Deutschland
produktsicherheit@kolibri360.de